Books are to be returned on or before
the last date below.

Chemistry and Medicines
An Introductory Text

Chemistry and Medicines
An Introductory Text

James R Hanson
Department of Chemistry, University of Sussex, Brighton, UK

RSCPublishing

ISBN-10: 0-85404-645-3
ISBN-13: 978-0-85404-645-4

A catalogue record for this book is available from the British Library

Published by The Royal Society of Chemistry,
Thomas Graham House, Science Park, Milton Road,
Cambridge CB4 0WF, UK

Registered Charity Number 207890

For further information see our web site at www.rsc.org

Typeset by Macmillan India Ltd, Bangalore, India
Printed by Henry Lings Ltd, Dorchester, Dorset, UK

Preface

The aim of this book is to provide a brief introduction to medicinal chemistry for final year chemistry and biochemistry undergraduates and for chemistry postgraduates. Many final year university courses include a 16–20 lecture course on medicinal chemistry as part of a programme of options. This book is based on short lecture courses, which have been given to these groups of students and on summer schools at the University of Sussex.

Medicinal chemistry utilizes aspects of bio-organic chemistry, organic synthetic methods, physical organic chemistry and organic reaction mechanisms. Many chemists are employed by the pharmaceutical industry in the design and synthesis of new drugs and in establishing structure:activity relationships. Chemists play a major role in identifying the metabolites of drugs, in formulating medicines and in their analysis. It is hoped that this book will provide a broad, short introduction to the chemistry of medicines for chemists starting work in these areas. The first two chapters provide an introduction to the subject arid describe some of the general chemical features that can affect a drug between the site of administration and the site of action. The remaining chapters are concerned with the role of medicinal chemistry in treating specific therapeutic targets. This approach has been adopted in this introductory text because the students may be aware of these diseases from their general knowledge. The development of drugs to treat these diseases illustrate many of the general strategies of medicinal chemistry.

I wish to thank the Royal Society of Chemistry and their referees for their help in the production of this book.

James R.Hanson
University of Sussex

Contents

Introduction

1.1 AIMS

The aims of this chapter are to introduce the structure of medicinal chemistry and to show how the subject has developed. By the end of this chapter you should know the salient points concerning:

- the basis for the classification of drugs;
- the targets for the medicinal chemist;
- the stages in the development of a drug; and
- the history of medicinal chemistry.

Medicinal chemistry is concerned with the chemistry of compounds that have a beneficial effect on a disease. Its objective is to enhance this beneficial therapeutic effect of a compound by modifying its structure and to remove unwanted side effects through an understanding of the chemistry by which the compound exerts its biological effects. Once a lead substance with a useful biological effect has been discovered, the medicinal chemist must undertake a series of structural variations in order to establish a pattern of structure: activity relationships leading to an enhancement of the useful biological effect. While the medicinal chemist is primarily concerned with the synthesis of new therapeutic agents, there must be a considerable interaction with other disciplines; medicine and biology for the description of the disease state and the development of the bio-assay, pharmacology for the definition of the site of action and pharmacy for the delivery of the compound to the living system.

1.2 THE CLASSIFICATION OF DRUGS

Drugs may be classified in a number of ways. One way is in terms of the nature of the disease that they are being used to treat. Thus there are compounds that are used to treat infectious diseases. Second, there are compounds that are used to treat cancers and third, non-infectious diseases. The chemistry of infectious diseases is concerned with the development of drugs to injure an invading organism with minimal injury to the host. The targets are often the differences between the viral, bacterial, fungal or parasitic cells and those of man. This is the area of the antibiotics. The medicinal chemistry that is used in the treatment of cancers involves the use of drugs to destroy an aberrant cell within the host. The targets of cancer chemotherapy are the differences between the rapidly proliferating cancer cells and normal cells. Cancer chemotherapy is often used in conjunction with other forms of treatment such as radiotherapy. The chemistry of non-infectious diseases involves a study of the selective action of a drug on one cell or receptor in the host. In some cases the drugs are developed as mimics of natural hormones.

The drugs that are used in the treatment of non-infectious diseases can be further sub-divided in terms of their targets. There are a group of hormones known as neurotransmitters that are formed at nerve endings and convey the consequences of a nerve impulse to a receptor or an effector cell. There are drugs, which affect primarily the neurotransmitters in the central nervous system including the brain and the spinal cord. These include the psychotropic agents, the anti-depressants, hypnotics and analgesics. There are those agents, which affect the peripheral nervous system including the local anaesthetics. Another group of substances are those, which affect the circulatory system acting as anti-hypertensive and anti-thrombotic agents. These may interact with local hormones. These are a family of compounds that have a metabolic or endocrine target. These are compounds that are modifications of circulatory hormones that may be used as oral contraceptives or to correct a hormone deficiency. Finally there are those compounds, which target the immune system such as the immunosuppresive agents. However there is an overlap between these classes and a compound may show several types of biological activity.

The pattern of usage and the length of time over which a drug may be administered varies between the classes. Hence the extent to which side-effects can be tolerated varies quite widely. For example, an antibiotic may be given for a few days while a compound, which is given to correct a hormone deficiency or as an immunosuppresive agent may be administered for years.

1.3   TARGETS FOR THE MEDICINAL CHEMIST

1.3.1   Hormones as Targets

The body produces substances known as hormones, which regulate body functions. These can be circulatory hormones such as the steroid and peptide hormones. They are produced by one organ and are then transported to their target organ. Others, such as histamine are local hormones, which are produced by one cell and have their action on adjacent cells. These are sometimes known as autocoids. The third group are the neurotransmitters, which are formed and have their action at nerve endings. A fourth group are the 'second messengers'. These are compounds that are formed within a cell often as a result of an external stimulus via a *trans*-membrane protein. They control the function of various enzyme systems within the cell.

Many hormones and neurotransmitters exhibit their cell-signalling biological activity by binding to a receptor on a cell surface. The receptor may be part of a *trans*-membrane protein, which crosses the cell wall. This binding to a *trans*-membrane protein then initiates a sequence of events within the cell. Other hormones have to cross the cell wall and exert their biological activity by binding to nuclear receptors within the cell. This activates the nucleic acids and initiates the DNA–RNA–protein sequence of events.

The hormones are biosynthesized by a series of steps and once they have produced their biological effects, they are metabolized and excreted. The medicinal chemist may interact with this sequence in a number of ways. The chemist may synthesize the biological compound itself and use it to correct a deficiency or an agonist may be prepared. An agonist is a relative of the naturally occurring substance that also binds to the receptor and elicits the same biological effect. A partial agonist is a compound, which binds but does not elicit the full response. In contrast to this an antagonist binds to the receptor site but does not produce the biological effect. It may block the effect of an agonist. Often agonists and antagonists have quite similar structures for both have to bind to the receptor.

The enzymes which mediate the biosynthesis of the naturally-occurring compound may be inhibited by a drug. Hence the hormone will not be formed and its biological effect will not be observed. Many enzymes are regulated by a metabolite from a later stage in the biosynthetic pathway. This type of feedback regulation may be used to moderate the amount of biosynthesis that occurs. The release of a compound from storage may also be a regulatory step. When enzyme systems are targets for drugs, the binding may be of a competitive and reversible nature or it may be irreversible. Sometimes the product of the reaction of the enzyme with an artificial substrate may then react with the enzyme itself preventing the enzyme from catalyzing further transformations. This type of inhibition is known as suicide inhibition.

Once a hormone or neurotransmitter has completed its biological function, it may be metabolized and excreted or it may participate in the

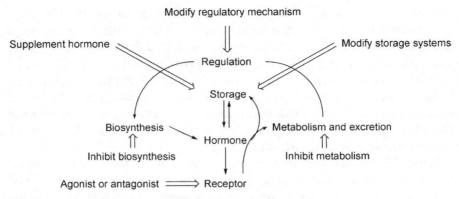

Scheme 1   *Hormonal targets for the medicinal chemist*

feedback regulatory mechanism associated with its formation. If these later metabolic steps are inhibited, the action of the hormone may be prolonged. The re-uptake of a neurotransmitter may form part of the regulatory mechanism. Interferance with the re-uptake mechanism may also prolong the action of the neurotransmitter. These targets are summarized in Scheme 1.

1.3.2   Cellular Structures as Targets

Cellular structures and cellular constituents provide another series of targets for the medicinal chemist. The structure and formation of the viral, bacterial or fungal cell wall provide a series of targets. These cell walls have different structures to those of man allowing for selective action. If a cell wall cannot be formed correctly, it may rupture and the cellular constituents may leak out.

A cell wall has ion-channels through which ions can pass. The ion-channels, which allow sodium, potassium, calcium or chloride ions to enter a cell, form targets. The presence of these ions in the cell affects processes such as muscle contraction.

The interaction of a cell-signalling substance with a receptor can initiate a further series of enzyme-catalysed events within the cell. These enzymes may be the target for drugs. The nucleic acids form further targets particularly in cancer chemotherapy. Interferance with the bio-synthesis of nucleic acids, their translation and replication, each form targets for drugs. Thus there are a plethora of potential targets for drug action. A crucial stage in the combat of a disease is the selection of an appropriate target. The selection of the target determines the bio-assay for a drug.

1.4 THE STAGES IN THE DEVELOPMENT OF A DRUG

The first stage in the development of a drug involves the establishment of a reliable bio-assay. This may be an antibiotic screen or a screen against a particular tumour cell line. It may involve enzyme or receptor assays. These days whole animal tests are rare in the primary screens although in previous years, they have played an important part. Nevertheless because of the complexity of biological systems, there are still situations in which whole animal tests have to be used to obtain information on a potential drug. Modern enzyme or receptor screens can be very rapid and have a high throughput allowing small amounts of many thousands of compounds to be screened within weeks. This in turn has led to new synthetic methodologies in organic chemistry such as combinatorial synthesis in order to generate a suitable range of compounds for testing.

There are a number of sources of lead compounds. These may be obtained by screening natural products particularly from plants that have been used in folk medicine. The lead compound may arise from random screening or from the clinical observation of a side effect of an existing drug. The rational design of a lead compound based on the structural modification of a hormone or an active site model, is an intellectually satisfying approach.

Once a lead compound has been identified, there is a progressive structural modification to enhance the activity and to identify the contribution of electronic and steric factors to the biological activity. This can be with the aid of computer based molecular-modelling techniques. The establishment of quantitative structure: activity relationships (QSAR) can lead to the identification of a part of the molecule that is responsible for the activity, the pharmacophore.

Drug metabolism studies and pharmacokinetic studies then follow. The identification of the metabolites of a drug can involve the medicinal chemist in the preparation of labelled material. Once a compound is under serious consideration as a drug candidate, animal toxicity studies are undertaken. On the chemical side the formulation and development of a manufacturing route and appropriate analytical methods have to be undertaken. Clinical studies then follow. The phase I trials involve healthy volunteers and aim to establish the acceptability of a compound in man and obtain some pharmacokinetic data, phase II trials are with a limited number of patients and aim to establish the efficacy of the drug. This includes proof of the principle underlying the activity. Finally phase III large-scale trials are used to establish the efficacy of a drug compared to its rivals. Evidence has to be obtained concerning the safety of the drug and any contra-indications for its use. During any one of these phases development may be stopped if toxicity is detected. The submissions to the drug safety and drug registration committees then follow such as the Medicines and Health Care Products Regulatory Authority and the U.S. Food and Drug Administration. Finally if a compound is to be prescribed in the National Health Service, its effectiveness has to be assessed and a recommendation made by the National Institute of Clinical Excellence (NICE). The current time scale between the initial programme and the release of a compound for use may be between 10 and 15 years and the cost may be of the order of £500 million pounds. The need for patent protection in these circumstances is obvious. This sequence of events is summarized in Scheme 2.

1.5   THE SYNTHESIS OF A DRUG

Synthetic organic chemistry is one of the corner stones of medicinal chemistry. There are a number of criteria by which a synthesis may be evaluated in a medicinal chemistry context. A convergent synthesis rather than a linear synthesis (Scheme 3) has significant advantages. Not only are there benefits in terms of yield but there is increased flexibility. Structural variation is possible in one arm of the synthesis while keeping the other constant and vice versa. This enables structure: activity studies to be made more easily. Metabolic studies require the preparation of labelled material. It is necessary to explore not only what happens to the intact drug in the body but also to trace metabolic fragments. A convergent synthesis makes labelling different parts of the molecule easier. If a chiral centre is created in the drug then the synthesis should not only be stereospecific but also enantiospecific. The targets for most drugs are chiral. Although the required biological activity may reside in one enantiomer, the other enantiomer may have different, potentially toxic properties. A racemic mixture has to be avoided. Biotransformations involving a chiral enzymatic step have an increasingly important role to play in the preparation of a single enantiomer.

While retrosynthetic analysis must play an important role in the design of a synthetic scheme, economic considerations in terms of the availability of starting materials play an equal part. The art of synthesis in a medicinal chemistry context lies in the combination of retrosynthetic analysis with the identification of readily available basic building blocks in the target structure. The dissection of a structure into its basic building blocks can also be a useful way of remembering the structure of drugs.

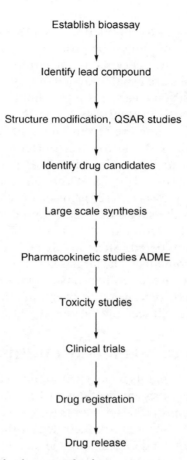

Establish bioassay

Identify lead compound

Structure modification, QSAR studies

Identify drug candidates

Large scale synthesis

Pharmacokinetic studies ADME

Toxicity studies

Clinical trials

Drug registration

Drug release

Scheme 2   *Stages in the development of a drug*

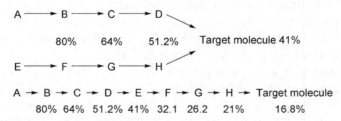

A ⟶ B ⟶ C ⟶ D

80%   64%   51.2%   Target molecule 41%

E ⟶ F ⟶ G ⟶ H

A → B → C → D → E → F → G → H → Target molecule
80% 64% 51.2% 41% 32.1 26.2 21%   16.8%

Scheme 3   *The effect of an 80% yield at each step on a convergent and a linear synthesis*

In recent years high throughput enzyme and receptor screens have been developed which require large numbers of small samples for testing. This has placed considerable demands on synthetic chemists who have responded by introducing automated combinatorial methods of synthesis. The object of a combinatorial synthesis is to maximize the number of compounds that might be produced by simple combinations of starting materials and reagents to generate a library of related structures. The reactions are usually carried out by attaching the starting materials to a solid phase such as a resin by a linker. In a simple example there might be two starting materials, A and B, which are attached to separate sets of resins. These are mixed and split into two and each is then reacted with either C or D to give four compounds. If these are mixed and split again to be combined with E and F, there are eight possible combinations ($2 \times 2 \times 2$). Although this is a small library, larger libraries (*e.g.* $5 \times 5 \times 5 = 125$) can be developed quite rapidly.

Once a drug reaches the stages of toxicity and clinical trials, relatively large quantities will be required. A number of additional chemical features have to be considered. Steps involving low yields and difficult separations must be eliminated. The large-scale use of hazardous reagents or those, which produce toxic residues have to be avoided. A laboratory synthesis may need to be redesigned to overcome these problems. In the subsequent chapters we will see how various syntheses fulfil these criteria.

1.6   THE HISTORY OF MEDICINAL CHEMISTRY

There is a long history of plants being used to treat various diseases. They figure in the records of early civilisations in Babylon, Egypt, India and China. The therapeutic properties of plants were described by the Ancient Greeks and by the Romans and are recorded in the writings of Hippocrates, Dioscorides, Pliny and Galenus. Some metals and metal salts were also used at this time. In the Middle Ages various 'Materia Medica' and pharmacopeas brought together traditional uses of plants. The herbals of John Gerard (1596), John Parkinson (1640) and Nicolas Culpeper (1649) provide an insight into this widespread use of herbs. Exploration in the seventeenth and eighteenth centuries led to the addition of a number of useful tropical plants to those of European origin.

The nineteenth century saw the beginnings of modern organic chemistry and consequently of medicinal chemistry. Their development is intertwined. The isolation of a number of alkaloids including morphine (1805), quinine (1823) and atropine (1834) from crude medicinal plant extracts was part of the analytical effort to standardize drug preparations and overcome fraud.

General anaesthetics were introduced in surgery from 1842 onwards (diethyl ether (1842), nitrous oxide (1845) and chloroform (1847)). Antiseptics such as iodine (1839) and phenol (1860) also made an important contribution to the success of surgery. The hypnotic activity of chloral (trichloroethanal) (1869) was also reported.

Many of the developments after the 1860s arose from the synthesis of compounds specifically for their medicinal action. Although the use of willow bark as a pain-killer was known to the herbalists, the analgesic activity of its constituent salicin **1.1** and of salicylic acid **1.2** were developed in the 1860s and 1870s. p-Hydroxyacetanilide **1.4** (paracetamol) and phenacetin **1.5** (1886) were also recognized as pain-killers. Acetylation of salicylic acid to reduce its deleterious effect on the stomach led to the introduction of aspirin **1.3** in 1899. However its mode of action was not established until 1971.

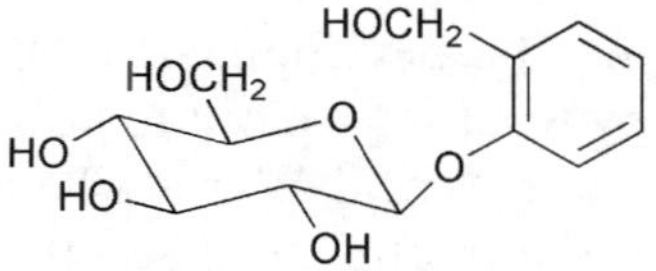

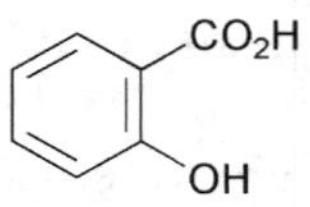

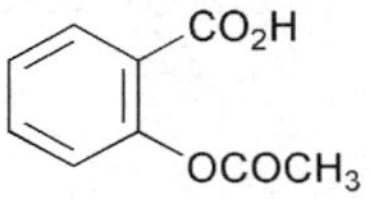

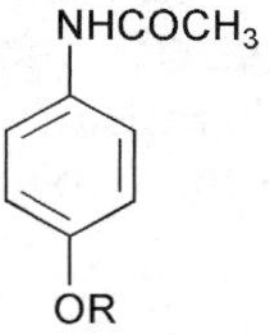

The local anaesthetic action of cocaine was reported in 1884 although its structure was not known at the time.

Various modifications of the dialkylamino esters of aromatic acids modelled on part of the structure of cocaine **1.6** led to benzocaine (1892) and procaine **1.7** (1905). The barbiturates, veronal (1903) and phenobarbital **1.8** (1911) were introduced as sleeping tablets.

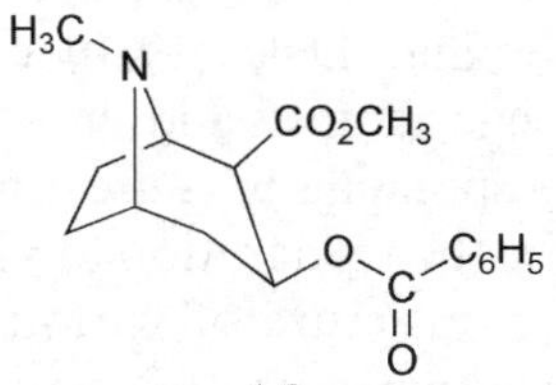

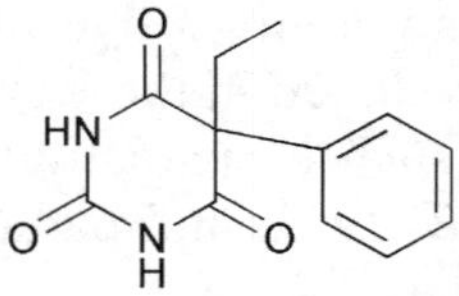

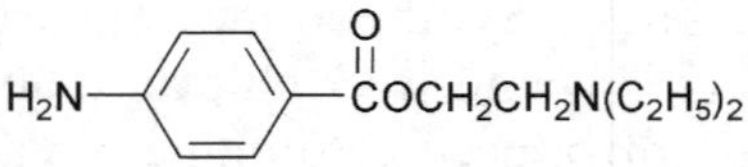

Once ideas of chemical structure were formulated in the mid-nineteenth century, the first theories of the relationships between chemical structure and biological activity began to emerge. Thus Crum-Brown and Fraser (1869) noted that a 'relationship exists between the physiological action of a substance and its chemical composition' leading to the idea that cells can respond to the signals from specific molecules. On the basis of observations that certain dyes selectively stained micro-organisms, Ehrlich in the 1890s put forward the idea that there were specific receptors for biologically active compounds – 'lock and key' relationships. This led to the examination in 1904 of dyestuffs such as trypan red for the treatment of trypanosomiasis and the development (1907) of salvarsan **1.9** for the treatment of syphilis by what turned out to be a false structural analogy (see Chapter 6). In the First World War acriflavine and proflavine **1.10** dyestuffs were used for the treatment of sepsis in wounds. The work of Meyer and Overton (1899–1901) to relate a physical property (the oil: water distribution co-efficient) to biological activity (anaesthesia) were the first rudimentary QSAR. Another quantitative measurement that was made was the chemotherapeutic index, which was the ratio of the minimum curative dose to the maximum tolerated dose (CD50/LD50).

1.9   1.10   1.11

The action of acetylcholine on nerve tissue had been recognized in the late nineteenth century. Barger and Dale (1910) examined the response of various tissues to acetylcholine agonists and showed that there were different receptor sub-types; some responding to muscarine and others to nicotine.

The 1920s and 1930s saw the recognition of vitamin deficiency diseases and the elucidation of the structure of various vitamins. It was also a period in which there was exposure of many Europeans to tropical diseases. The iodinated quinolines such as entero-vioform **1.11** were introduced to combat amoebic dysentary and complex dyestuff derivatives such as suramin and germanin were developed in the 1920s to treat sleeping sickness. Synthetic anti-malarials such as pamaquine (1926), mepacrine (1932) and later chloroquine **1.12** (1943) and paludrine **1.13** (1946) were introduced as quinine replacements.

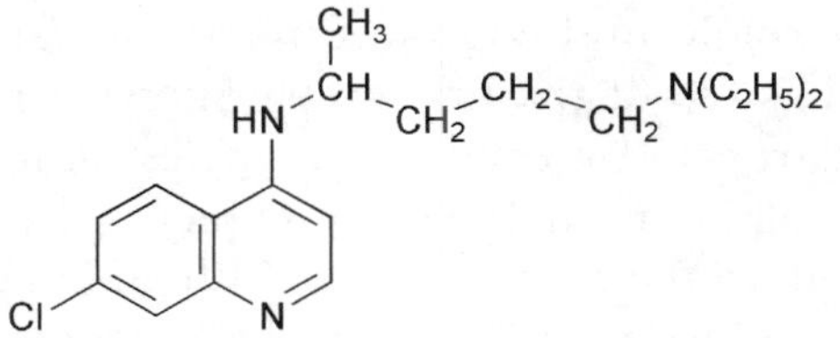

1.12

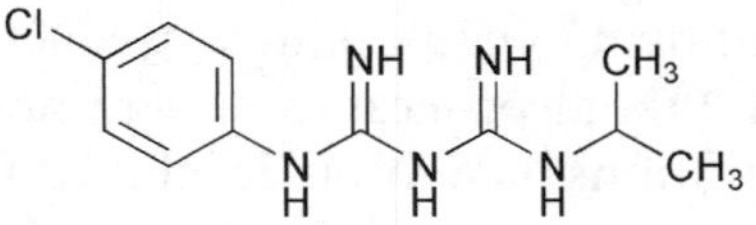

1.13

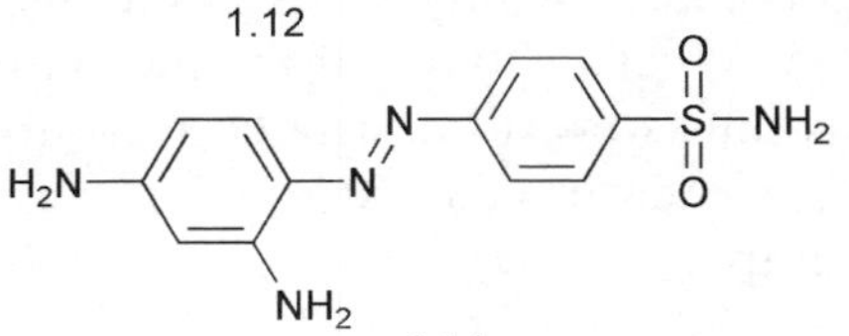

1.14

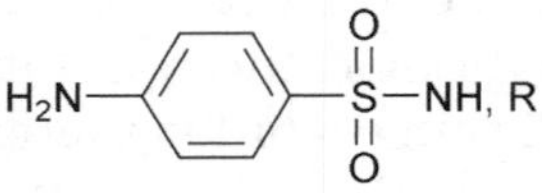

1.15

In 1935 Domagk observed the anti-bacterial action of the sulfonamide dyestuff, prontosil red **1.14**, from which the important family of sulfonamide **1.15** anti-bacterial agents were developed. The activity of these compounds as inhibitors of folic acid biosynthesis was rationalized by Woods (1940) as anti-metabolites of p-aminobenzoic acid. With the onset of the Second World War, there was a need for new antibiotics. In 1929 Fleming had observed that a strain of *Penicillium notatum* inhibited the growth of a *Staphylococcus*. In 1940–1941 Chain, Florey and Heaton isolated benzylpenicillin **1.16**. After considerable chemical work, the β-lactam structure for the penicillins was established. The relatively easy bio-assays for anti-bacterial and anti-fungal activity led to the isolation of a number of antibiotics including streptomycin (1944), chloramphenicol (1949) and the tetracyclines such as aureomycin **1.17** (1949).

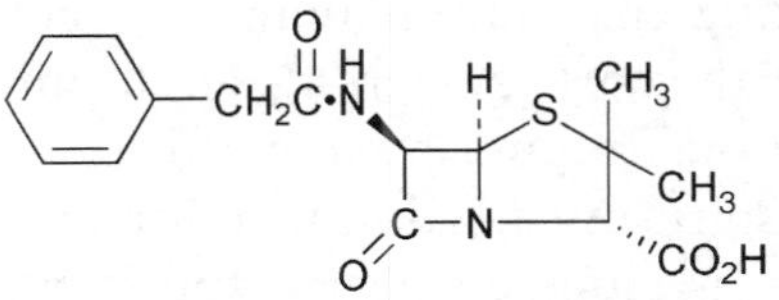

1.16

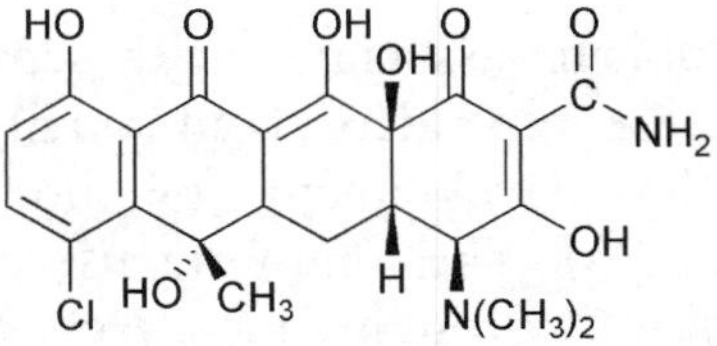

1.17

Several different aspects of medicinal chemistry developed in parallel through the second half of the twentieth century. Although they did not develop independantly, it is easier to follow their progression by considering them separately.

The structures of the steroid hormones were established in the 1930s and 1940s. The discovery in 1949 of the beneficial effect of cortisone **1.18** in alleviating the inflammation associated with rheumatism provided the stimulus for synthetic activity in this area. A number of anti-inflammatory semi-synthetic corticosteroids such as prednisolone, betamethasone **1.19** and triamcinolone became available in the late 1950s and 1960s.

Animal experiments to develop steroidal oral contraceptives were carried out before the Second World War but the first preparations (*e.g.* Enovid®) containing a synthetic estrogen, for example ethynylestradiol **1.20** and progestogen were not available until 1959. Subsequent preparations have been developed to reduce the estrogen level. Mifepristone **1.21**, which is an anti-progestogen and forms the basis of the 'morning-after pill', was introduced in 1985. Whereas many of the medicines that had been developed prior to this time were administered for only short periods of time, this was not true of the steroids and concerns developed over the effects of long-term therapy.

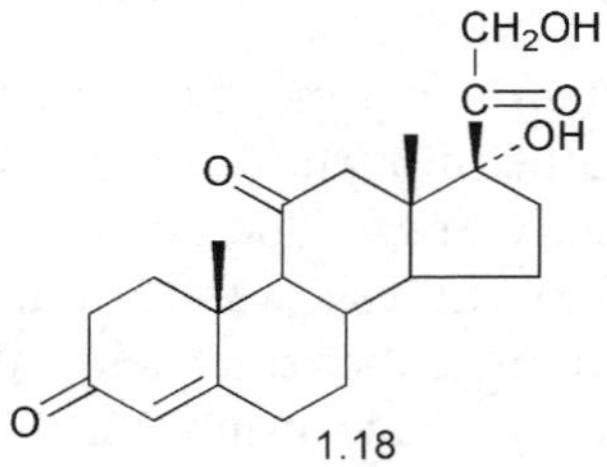

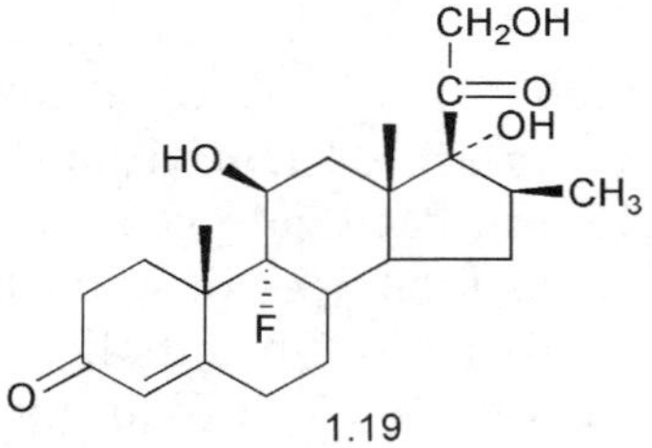

Problems associated with separating the anti-inflammatory activity from the mineralcorticoid activity of the cortical steroids led to interest in the development of non-steroidal anti-inflammatory agents (NSAIDs). The long-term use of aspirin as a pain-killer for arthritic conditions brought side-effects such as stomach ulcers. Indomethacin and ibuprofen (nurofen®) **1.22** were introduced in 1965 and 1971 respectively as alternatives.

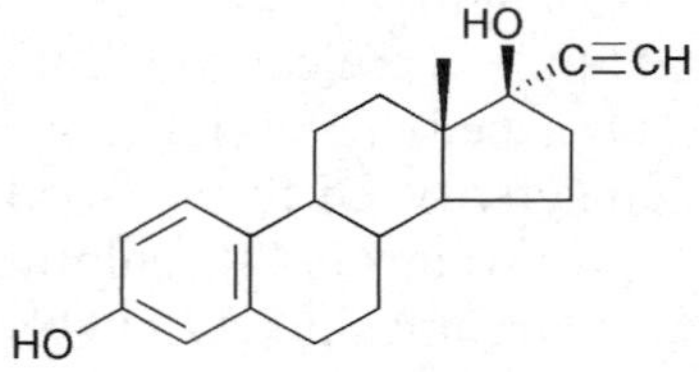

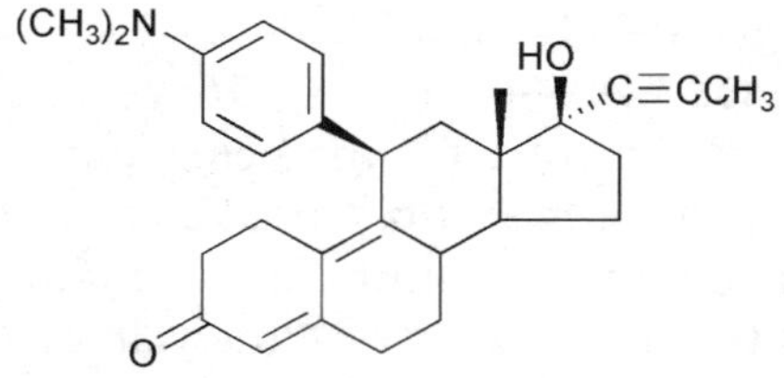

During the 1960s the prostaglandin hormones were implicated in inflammation and in the protection of the stomach against ulcers. In 1971 aspirin was shown to inhibit the biosynthesis of the prostaglandins from arachidonic acid by the enzyme system, cyclo-oxygenase. The subsequent realization that there were several forms of cyclo-oxygenase provided the framework for developing selective non-steroidal anti-inflammatory agents that only targeted some of the multiple activities of the prostaglandins. One result was the introduction in 1999 of celecoxib (Celebrex®) **1.23** and rofecoxib (Vioxx®) as selective cyclo-oxygenase (COX-2) inhibitors. Recently cardiovascular side effects of these compounds have begun to emerge and Vioxx® has been withdrawn.

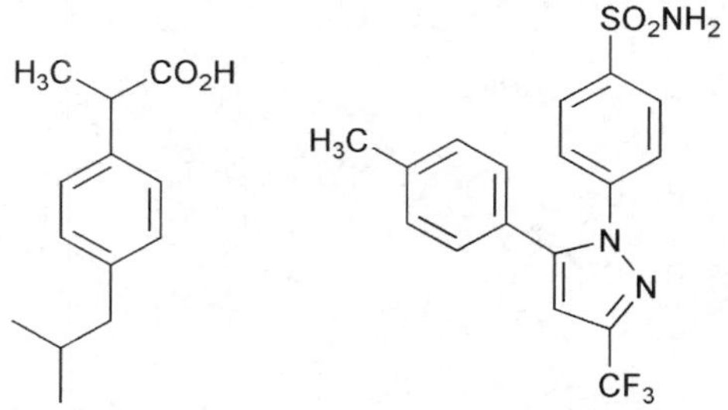

1.22   1.23

A number of developments took place in the 1960s, which changed medicinal chemistry. It was found that a drug, thalidomide **1.24**, which had been introduced as a sedative, when used by pregnant women, led to the birth of deformed children. The consequences of this teratogenic effect brought about a major tightening of the regulations regarding drug registration and the safety of medicines. Unfortunately there was some tardiness in the recognition of this side-effect. Second in 1964 Hansch published correlations between substituent effects (Hammett parameters) and the biological activity of some aromatic compounds. These QSAR began to provide a framework for the systematic development of drugs and for decisions to be made in the planning of a research programme.

The logical development during the 1960s of histamine antagonists for the treatment of peptic ulcers led to cimetidine **1.25** (1976) and then ranitidine (1981). The reasoning behind this work had a major impact on the development of medicinal chemistry.

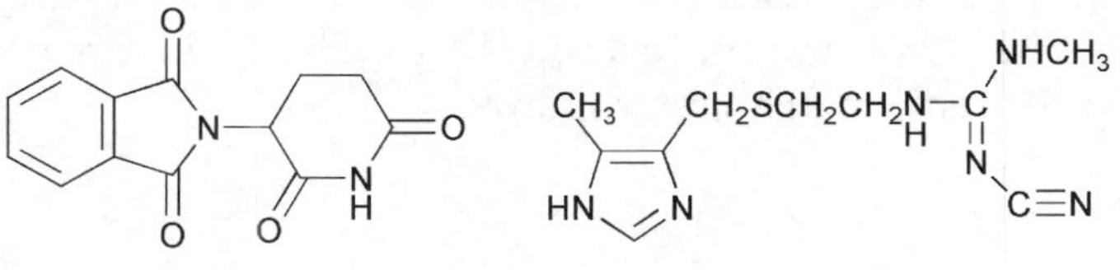

1.24   1.25

Adrenalin (epinephrine) **1.26** was the first substance to be recognized as a hormone (1901). The adrenergic receptors were divided in the α- and β-receptors by Ahlquist in 1948 based on their responses to selective agonists, *e.g.* isoprenaline **1.27**. The β-receptors were subsequently subdivided by Lands. This, together with an understanding of the metabolism of adrenalin (epinephrine) led to the discovery of salbutamol **1.28** (1967) as a selective β_2 agonist in the treatment of asthma.

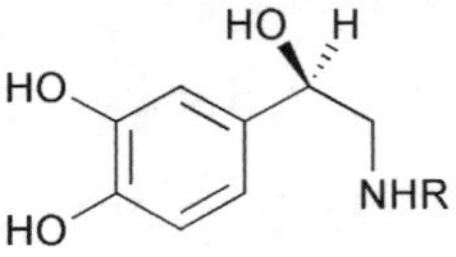

1.26 R = CH$_3$

1.27 R = CH$\begin{smallmatrix}CH_3\\CH_3\end{smallmatrix}$

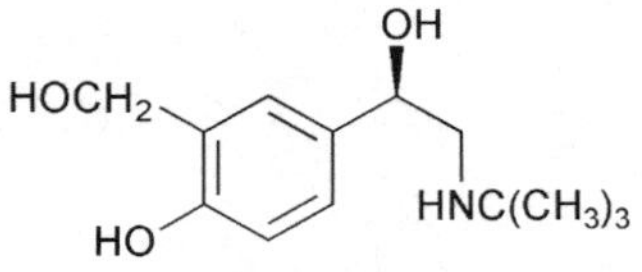

1.28

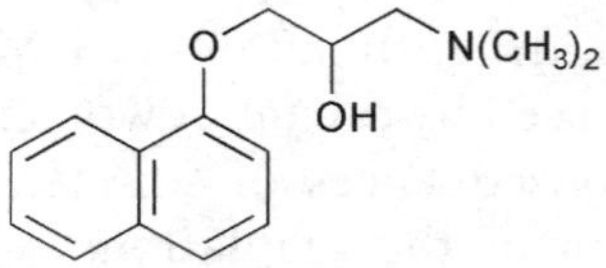

1.29

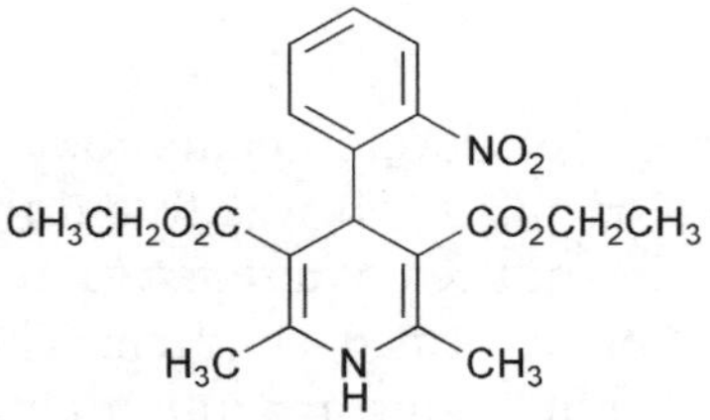

1.30

The development of drugs such as propranolol **1.29** (1964) and atenolol (1970), which blocked the β_1 receptors in cardiac muscle, was a major advance in the treatment of heart disease. Another important group of drugs with cardiovascular properties that were developed in the 1960s were those that block the movement of calcium ions through ion-channels. These were dihydropyridines such as nifedipine **1.30** (1967). Angiotension converting enzyme (ACE) inhibitors such as captopril **1.31** (1977) and enalapril (1984) are further valuable anti-hypertensive agents. The association of high cholesterol levels with circulatory diseases led to the development of cholesterol biosynthesis inhibitors. A major family of drugs with this activity are the statins exemplified by lovastatin **1.32** (1987) and simvastatin (1988). The statins are now widely prescribed for reducing cholesterol levels.

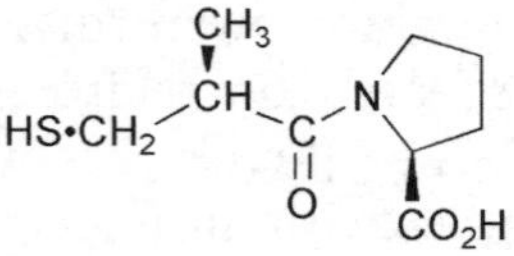

1.31

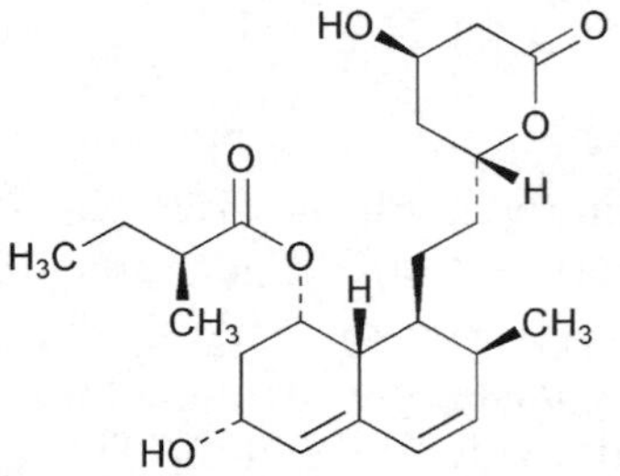

1.32

Medicinal chemistry has revolutionized the treatment of mental disease during the second half of the twentieth century. An increasing understanding of the role of various neurotransmitters in the brain has played an important part in this. A number of anti-depressants and antipsychotic agents were developed in the 1950s including the phenothiazine, chlorpromazine **1.33** (1952), the tricyclic compounds such as imipramine (1957) and the butyrophenones such as haloperidol (1958). The benzodiazepine tranquillizers such as librium **1.34** (1960) and valium (1963) were the forerunners of a large family of drugs.

1.33   1.34

The discovery in 1950 that a dopamine deficiency was associated with the neurodegenerative disease known as Parkinson's disease, led to various strategies to overcome this dopamine deficiency. Unfortunately just administering dopamine was unsuccessful because it did not reach the brain. However a successful treatment made use of its biosynthetic precursor, L-DOPA **1.35** (1961). DOPA decarboxylase inhibitors such as carbidopa were developed to increase the amount of L-DOPA reaching the brain by preventing its decarboxylation before it reached the blood:-brain barrier. Dopamine agonists such as pergolide (1988) and ropinirole **1.36** (1996) and inhibitors of dopamine metabolism such as tolcapone (1997) have provided other methods of treatment. Whereas diminished dopamine levels have been associated with neurodegenerative diseases, excessive responses to dopamine and other neurotransmitters are associated with different conditions. Some aspects of depression have been associated with reduced levels of the neurotransmitter, serotonin. This has culminated in the development of selective serotonin reuptake inhibitors such as fluoxetine (Prozac®) **1.37**, paroxetine (Seroxat®) **1.38**, sertindole (1996) and olanzapine (1996).

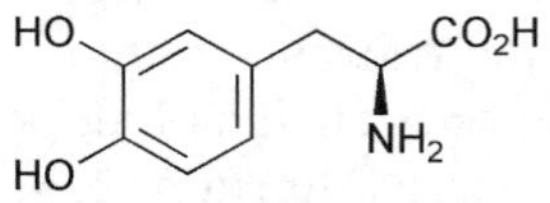

1.35 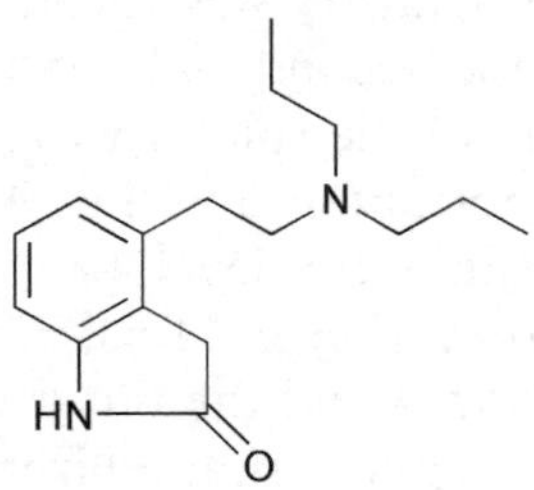   1.36

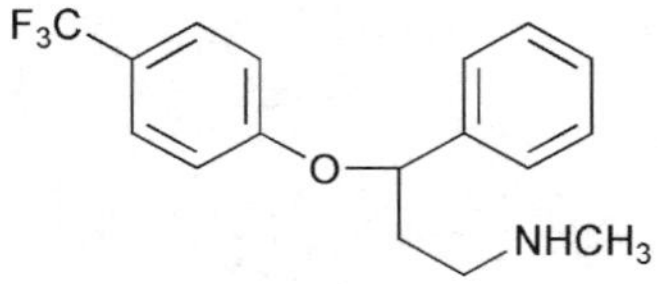

1.37 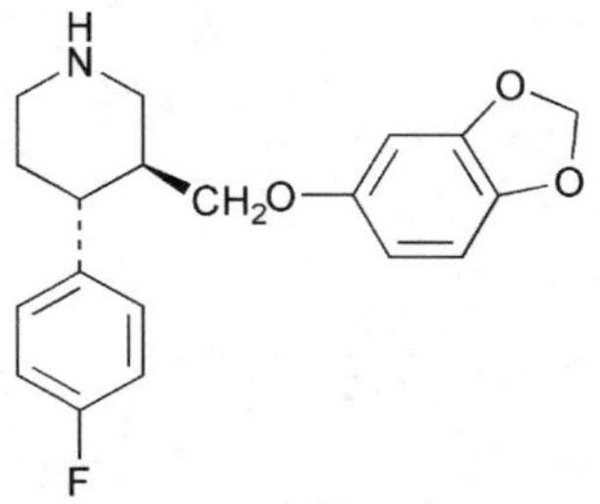   1.38

A major problem asociated with the treatment of many infectious diseases has been the development of resistant organisms. This has been found with viruses, bacteria and with parasitic organisms such as malaria. Strains of *Staphylococcus aureus* that were resistant to the natural penicillins were already starting to appear by the late 1940s. The penicillins that were used in the late 1940s and 1950s also had problems of stability associated with them. A significant step forward came in 1959 when methods for the commercial isolation of the 6-amino-penicillanic acid **1.39** core of the penicillins were developed. This permitted the synthesis of a range of semi-synthetic penicillins with enhanced stability and activity. Methicillin, ampicillin and amoxycillin **1.40** were introduced in 1960, 1961 and 1964 respectively. The related cephalosporin β-lactam antibiotics, cephaloridine, cephaloxin and cefaclor were introduced in 1964, 1967 and 1974 respectively. The development of resistant strains of bacteria possessing β-lactamases that degrade the penicillins, has become a serious problem. The combination of a β-lactamse inhibitor, clavulanic acid **1.41** (1976) with a penicillin, amoxycillin, in an antibiotic preparation known as Augmentin®, was one useful approach to the problem. However methicillin resistant strains of *Staphylococcus aureus* (MRSA) are an increasing problem. Although these may be combated with a different type of antibiotic, vancomycin, strains that are resistant even to this antibiotic are beginning to appear.

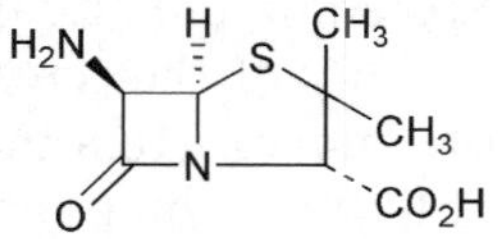

1.39

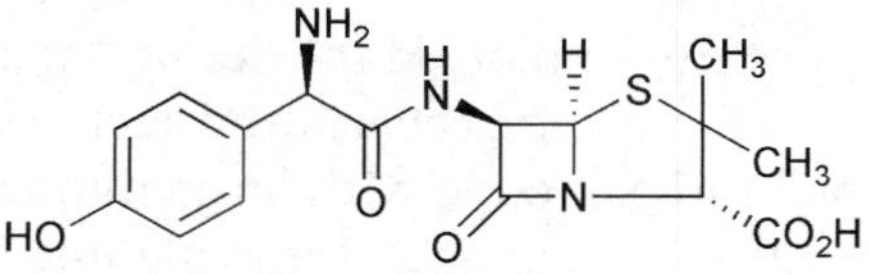

1.40

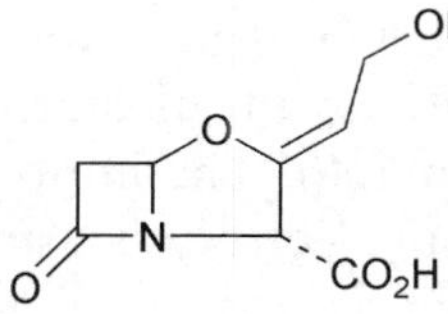

1.41

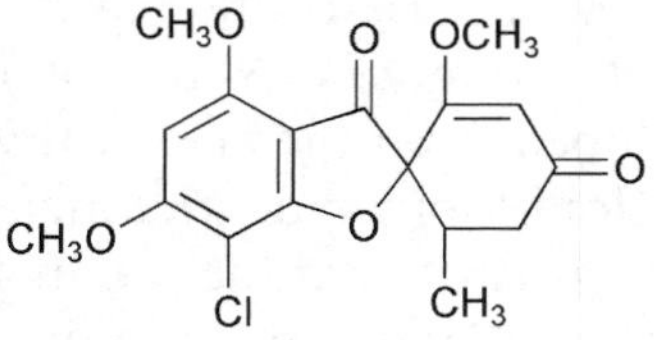

1.42

Fungal infections of man are mainly confined to the skin. A number of useful antifungal agents have been developed. The structure of the anti-fungal microbial metabolite, griseofulvin **1.42**, was established in 1952 and it was launched in 1959. Inhibition of the sterol component of the fungal cell wall has provided the basis of the action of a family of anti-fungal agents known as the azoles. These include micoconazole **1.43** (1972), ketoconazole (1980) and fluconazole (1988).

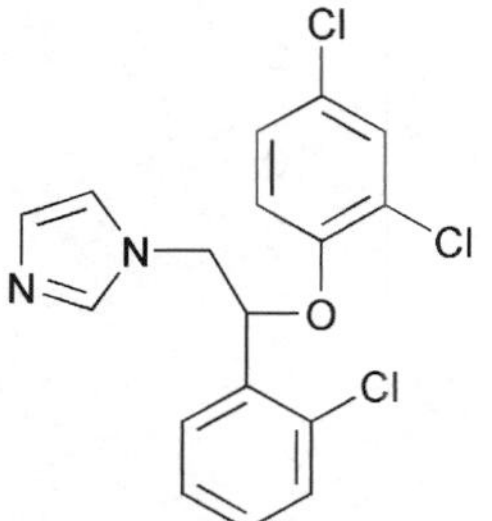

1.43

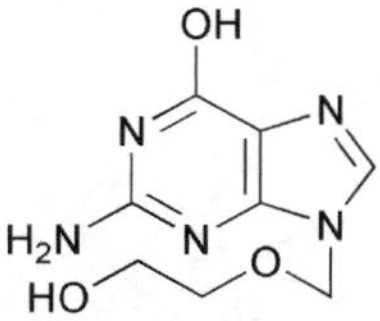

1.44

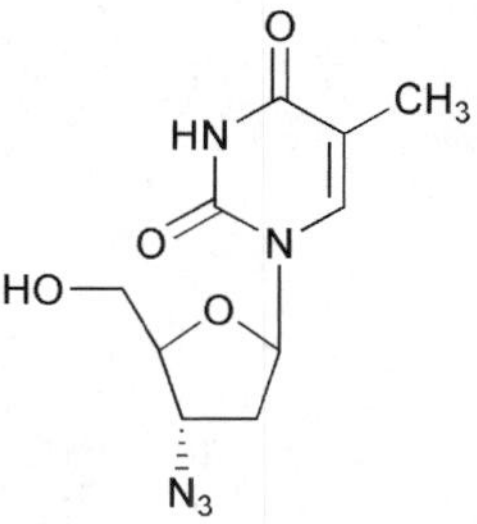

1.45

Whereas the bio-assay of anti-bacterial and anti-fungal agents is relatively straightforward, a virus requires its host-cell in which to replicate. Hence the bio-assay of anti-viral agents was more difficult until cell culture techniques were developed. Anti-viral agents active against the herpes virus include acyclovir **1.44** (1981). The identification of the viral origin of HIV-AIDS in 1983 led to the introduction of azidothymidine **1.45** (AZT) in 1987 to combat this disease. More recently (1999) zanamavir (Relenza®) and oseltamivir (Tamiflu®) have been developed for the treatment of 'flu'.

A key change in the bio-assay of drugs in the latter part of the twentieth century involved the development of receptor and enzyme bio-assays and the use of cell culture techniques. Many of the screens are very rapid and can cope with large numbers of samples. High through-put screening has changed the scale and rate at which compounds are produced for bio-assay.

The development of cancer chemotherapy has reflected this shift in screening from the use of animal models towards cell-lines associated with particular tumours. Many of the earlier drugs were alkylating agents developed from the chemical warfare agent, mustard gas. These included cyclophosphamide and melphalen **1.46**. The important obser-vation, made in 1969, that the products from electrolysis using platinum electrodes, slowed down the growth of bacteria, led to the development of the anti-tumour drug, cis-platin **1.47**. Another approach involved blocking the biosynthesis of DNA using drugs known as anti-met-abolites, which was exemplified by methotrexate **1.48**. Natural products, such as the Vinca alkaloids, vincristine and vincaleukoblastine and more recently, taxol® (paclitaxel) from the yew tree are useful tumour inhib-itory agents.

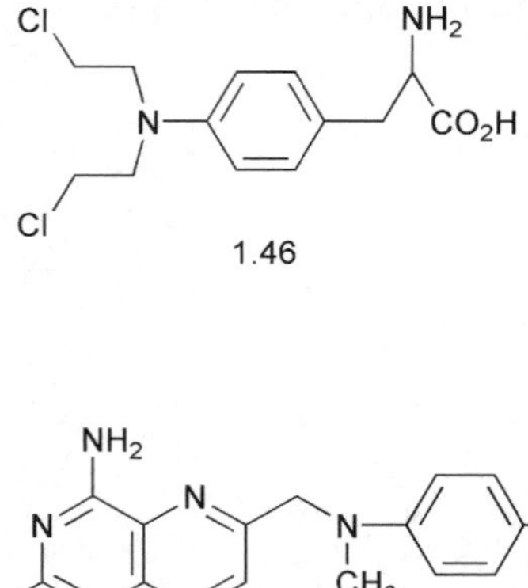

1.46 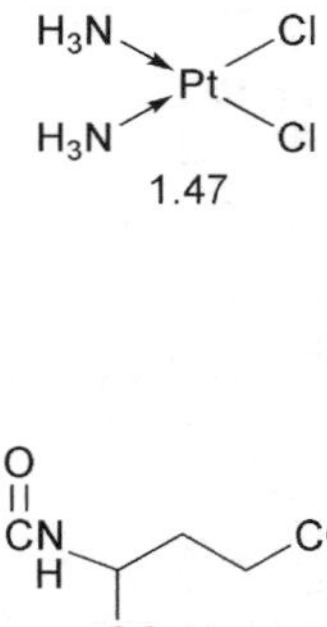   1.47

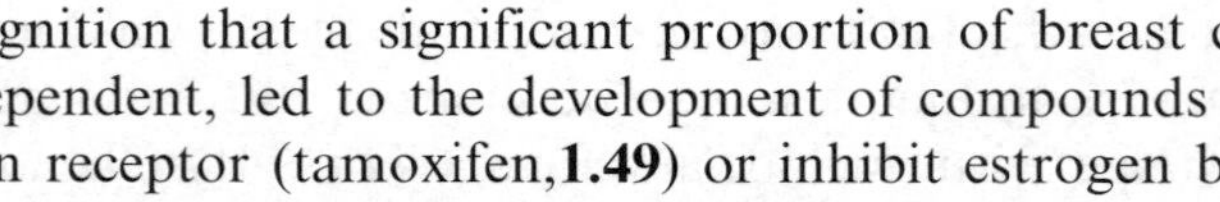

1.48

The recognition that a significant proportion of breast cancers are estrogen dependent, led to the development of compounds that target the estrogen receptor (tamoxifen,**1.49**) or inhibit estrogen biosynthesis (formestane, **1.50**, 1993; anastrazole, 1995). The use of monoclonal antibodies (*e.g.* herceptin) which recognize and specifically target par-ticular cancer cells and prevent them developing is a very important advance.

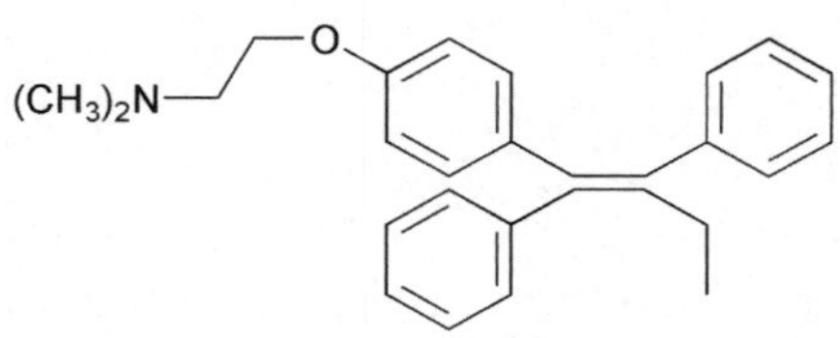

1.49 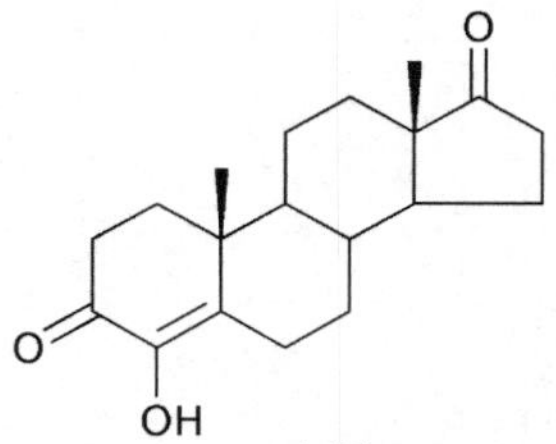   1.50

The impact of genomics on medicine and the recognition of genetic differences associated not only with specific diseases but also with the susceptibility to disease, is likely to lead to significant new treatments and refinements of older treatments. While many of these may involve the surgical introduction of particular cells, their ultimate success will retain a medicinal chemistry input. The diagnostic tests for many of these conditions also requires the skills of the medicinal chemist.

CHAPTER 2

General Principles of Medicinal Chemistry

2.1 AIMS

The aim of this chapter is to describe some of the general features that may affect the expression of biological activity by a drug. By the end of this chapter you should be aware of the general features that are associated with

- the absorption and distribution of a drug;
- the metabolic changes that may affect the excretion of a drug;
- some aspects of the metabolism of a drug that may lead to the development of toxicity; and
- the relationship between the biological activity and physico-chemical parameters such as pK, oil:water partition coefficients and steric factors.

2.2 ADMINISTRATION AND ABSORPTION

The route of administration, the absorption and the distribution, together with the metabolism and the excretion of a drug, can profoundly affect its use in man. There are four general routes by which a drug may be administered

  (i) orally, or less-commonly rectally, *via* the gastrointestinal tract;
  (ii) through the skin by topical (local) application;
  (iii) by injection (parenterally), which may be either intravenous or intramuscular; and
  (iv) by inhalation.

Each route has its particular advantages dependent on the condition that is being treated.

2.3 THE GASTROINTESTINAL TRACT

Administration *via* the gastrointestinal tract involves the drug crossing a number of barriers. A tablet may be dissolved in the mouth or a solution may be swallowed. The drug then enters the stomach *via* the oesophagus (see Scheme 1). The drug is subjected to digestive enzyme and bacterial changes. The strongly acidic nature of the stomach means that amines may be protonated while carboxylic acids will be in their unionized form. The intestine and colon have a pH that is closer to neutrality and hence there may be more free amine available, while even weak acids may be present as their salts. The pH of the stomach and the intestine varies during the day becoming less acidic after a meal. Hence, the timing of taking medicines can be important.

These physico-chemical changes clearly affect the water solubility of drugs and the ease with which they can cross lipid barriers into the blood stream. Once into the circulatory system, a drug will be transported to the liver that is a site of intense metabolic activity. The metabolic changes that

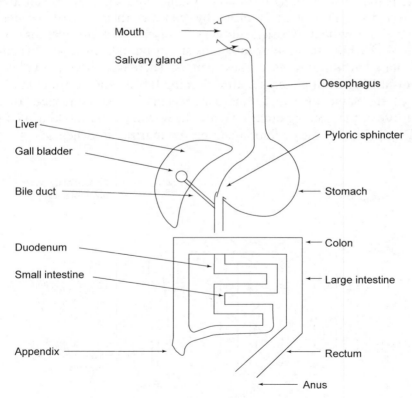

Scheme 1 *Schematic diagram of the gastrointestinal tract*

affect a drug in the liver may lead to its excretion prior to it reaching the target organ. This loss of material is known as the 'first pass loss'.

As the drug is translocated in the circulatory system, there are additional barriers to cross. A particularly important barrier is the 'blood brain' barrier, which prevents some compounds from reaching the brain. Ways of circumventing this barrier are important in developing drugs to treat various diseases of the brain. Structural alterations may also be made to a drug to prevent it reaching the brain and so reduce side effects arising from interactions within the brain. The membranes that a drug may have to cross typically comprise lipid (fatty) bilayers coating a protein through which pass various channels (see Scheme 2).

Whether the drug is in the intestine or in the circulatory system, it may not be present as free drug. It may be bound to food particles, to circulatory protein (albumin) or to nucleic acids and platelets. This means that a depot of a bound drug may be formed from which the drug is slowly released. The presence of such a depot may modify and reduce the biologically effective amount of a drug that is available at any one time. It may also prolong its action. Alcohol and some drugs can affect the bio-availability of other drugs by 'chasing' them out of the depot leading to a patient experiencing what is in its effect, an overdose. For example, barbiturate sleeping tablets exist substantially in depots. Alcohol can chase barbiturates out of these depots. There have been a number of deaths arising from patients taking sleeping tablets and consuming alcohol at the same time and thus experiencing the consequences of an overdose of the sleeping tablets. Hence the warnings about taking alcohol and medicines. These equilibria can be summarized in Scheme 3.

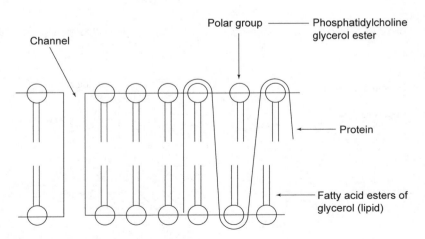

Scheme 2 *Schematic diagram of a membrane*

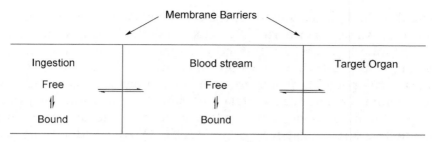

Scheme 3 *Equilibria between the free and bound drugs*

The combination of these physico-chemical and biological factors leads to the concept of the biological half-life of a drug in the system. However, it should be pointed out that a medicine might have a very different half-life in a healthy adult compared to one suffering from illness.

2.4 OTHER ROUTES OF ADMINISTRATION

When a compound is administered by topical application, the target is normally, but not always (*e.g.* a nicotine patch), local as with local anaesthetics. There is a fatty protective barrier on the skin that the substance has to traverse. The substance may therefore be applied in a solvent or as a cream that helps it to cross this barrier. A fat-solubilizing group such as ester may be chemically attached to the drug. This group may be removed subsequently by esterases within the body to reveal the active drug once it has crossed the barrier. The esterification of cortical steroids that are used in creams to alleviate skin conditions such as psoriasis, exemplifies this.

Intravenous injection provides a rapid route enabling a compound to enter the circulatory system without having to pass the liver and be subject to 'first pass loss'. While the onset of action from intravenous injection is quite rapid, that of intramuscular action is slower and the time-scale over which the drug acts is longer. Injection into the correct site is a skilled method of administration and its use is therefore restricted.

The final major method of administration, inhalation, is rapid and it is not restricted to compounds that are volatile. Indeed, a number of non-volatile compounds such as steroids that are used in the treatment of asthma are given by inhalation as aerosols. The advantage of inhalation in the treatment of asthma is the directness of action.

2.5 PHYSICO-CHEMICAL MEASUREMENTS

Various measurements on drugs can be made to shed light on the influence of physico-chemical factors on the bio-availability of a drug. Clearly the pK of a drug is going to reflect the extent to which a drug is ionized in the stomach and hence the conditions under which it may be absorbed. An oil:water partition co-efficient (octanol or less commonly decanol:water) can reflect transport across lipid barriers while the binding constant to bovine serum albumin can reflect the extent to which a compound is bound to circulatory protein. Measurement of these parameters and their correlation with the variation in biological activity for a series of compounds can be of value in establishing quantitative structure:activity relationships (QSAR) within a related group of compounds. This may be used to predict structural changes, which might then be made to enhance the biological activity.

2.6 FORMULATION

Although a typical tablet may weigh 200–300 mg, it may contain only 1 mg of active material. This poses problems of accurate dispensing and analysis. There must be an even distribution of the active drug in the tablet. It is no use having 49 tablets with no active constituent and 1 tablet containing 50 mg. The diluent may be lactose (milk sugar), dextrose or cellulose. The tablet may be held together by a binding agent such as acacia gum, sodium alginate or gelatin. Sometimes the tablet contains a lubricant such as a vegetable oil, magnesium stearate or polyethylene glycol and a wetting agent such as a detergent to facilitate the adminstration of the tablet. Some forms of starch swell on contact with water and therefore facilitate the disintegration of a tablet and its dissolution in water. The tablet may also contain colouring and flavouring matters. Formulation is an important aspect of medicinal chemistry. An active drug must be given in a form that is acceptable to a patient.

The rate at which a substance dissolves is proportional to the surface area of the particle. Small particles dissolve much more quickly than large ones. Different polymorphic crystalline forms of a drug may dissolve at a different rate and hence their overall absorption may be affected. This in turn can affect the biological activity of a drug. It is often the case that a patent may specify a particular polymorphic form of a drug.

Sometimes a tablet is coated with an 'enteric coating' such as cellulose acetate, which dissolves at a pH greater than 5.8. This prevents the tablet decomposing until after it has left the acidic regions of the stomach. The coating of a tablet may also provide a slow release form of the drug. Some drugs have a rapid onset of action and rapid excretion. Multiple doses over short time intervals may be required in order to maintain an effective concentration in the blood stream and yet avoid high toxic levels. This has been overcome by using a slow-release form of coating. For example, pseudoephedrine dries up the rhinitis in the symptoms of the common cold. However, too much taken at one time could have a serious effect on heart rate. Giving this drug in a slow-release capsule allows the effect on the rhinitis to last for upto 12 h while minimizing the harmful side effect.

Liquid formulations present a further series of medicinal chemistry problems. Many organic compounds are poorly soluble in water. Conversion of an acid or an amine to a salt may overcome this. The selection of the correct salt can be a matter of careful research. Many formulations contain alcohol or a syrup to overcome solubility problems. It is important to consider the shelf life of a liquid preparation. In a pharmacy and subsequently in the home, a bottle may be subject to photochemical reactions from sunlight and to autoxidation by air in a warm environment. These factors can contribute to drug stability. Many a medicine cupboard contains old bottles of drugs, which have decomposed or crystallized out. If you open an old bottle of aspirin, it is possible to smell the acetic acid arising from the decomposition of the acetylsalicylic acid. In some instances it is possible that the decomposition products may be toxic.

2.7 DRUG METABOLISM

A drug may undergo a metabolic change at any stage after ingestion, both prior to reaching its site of action and afterwards. Much of drug metabolism takes place in the intestine and in the liver. The metabolic changes may lead to the deactivation and excretion of a compound before it has had any effect, bringing about the first pass loss. The metabolism of a drug may also change its activity and bring about side effects. In other instances the metabolite may actually be the active species. In these cases there is a pro-drug:drug relationship.

As the drug passess through the system it is exposed to the digestive enzymes, which are often hydrolytic in their action. The bacterial action in the intestine is often reductive while the metabolism in the liver is often oxidative in character.

It is possible to distinguish two phases in the metabolism of a drug. In phase one, changes are brought about to the drug itself. Alcohols and aldehydes may be oxidized, hydroxyl or epoxide groups may be inserted, alkyl groups may be removed from nitrogen or oxygen and polar functional groups such as amino and hydroxyl groups unmasked by the hydrolysis of amides and esters. In phase two, a polar molecule such as glycine **2.1**, the sugar derivative, glucuronic acid **2.2**, a sulfate group **2.3** or the tripeptide glutathione **2.4** is linked to the drug by a process known as 'conjugation'. The overall effect of these changes is to increase the water solubility of the drug and to diminish its lipid solubility. This in turn can diminish the likelihood of it crossing various barriers and reaching its site of action. The addition of glutathione may remove a toxic metabolite.

$H_2NCH_2CO_2H$

2.1

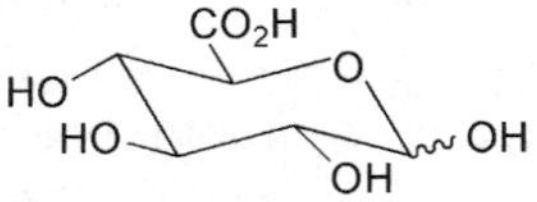

2.2

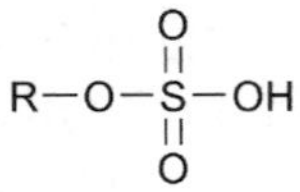

$R-O-\overset{\overset{O}{\|}}{\underset{\underset{O}{\|}}{S}}-OH$

2.3

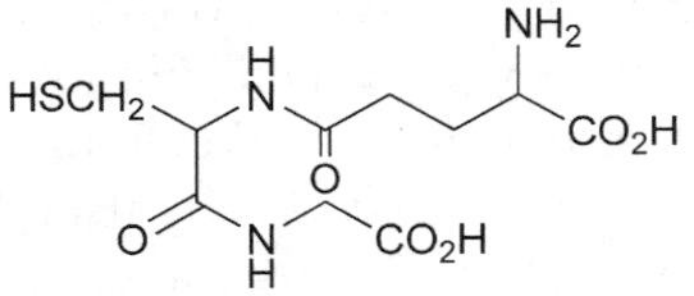

2.4

2.8   OXIDATION BY CYTOCHROME P$_{450}$S

There are a variety of oxidative changes that may occur to a drug. Many of these involve the insertion of oxygen using a cytochrome P$_{450}$-dependent enzyme. The cytochromes contain haem **2.5** as the co-enzyme. The oxygen is bound by the central iron and delivered to the substrate by a series of redox changes (see Scheme 4). There are a series of different cytochromes, CYP1A1, CYP1A2, CYP2A6, CYP2C9, CYP2C10, CYP2C18, CYP3A, CYP4A4 *etc.* in the liver. The cytochrome CYP3A4 is responsible for the oxidation of many drugs. Each of these, while containing the haem co-enzyme, has a range of different substrate specificities, hence the name 'mixed function oxidases'. The relative proportion of these enzymes can differ between individuals and thus there are variations in the metabolism of drugs. Some individuals may lack a particular cytochrome. Consequently, toxic metabolites may only become apparent after a drug has been in use with a large number of patients.

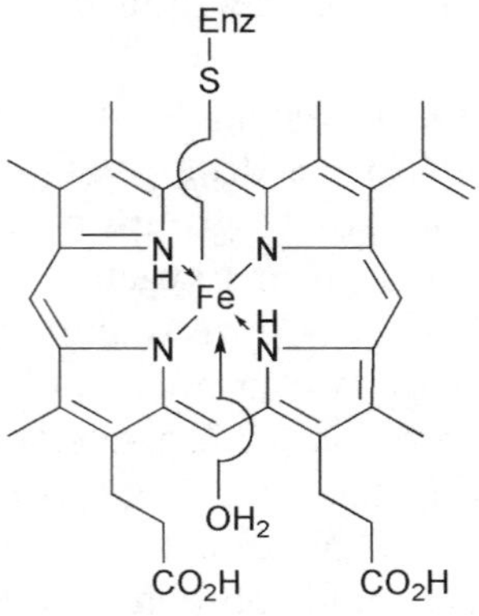

2.6

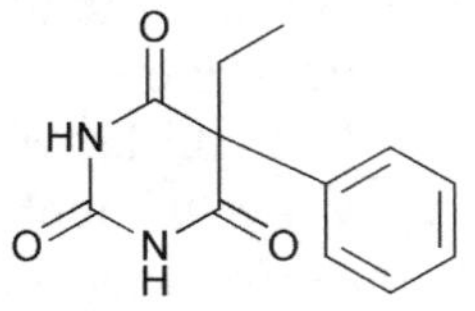

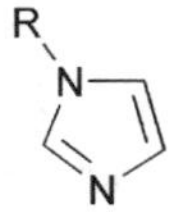

2.7

2.5

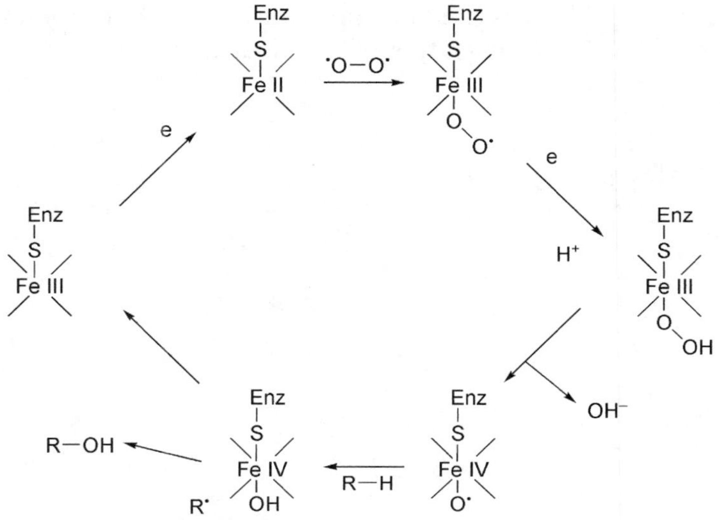

Enz
|
S
Fe II — ·O—O· → Enz
|
S
Fe III
|
O
O·

e

Enz
|
S
Fe III

H+

e

Enz
|
S
Fe III
|
O
OH

Enz
|
S
Fe IV
OH
R·

R—OH

Enz
|
S
Fe IV
O·

R—H

OH−

Scheme 4 *The role of iron in oxidation by a cytochrome P_{450}*

Some cytochrome P_{450}s in the liver are induced by specific drugs, *e.g.* phenobarbital **2.6** while others may be inhibited by drugs, *e.g.* by azoles **2.7**, which complex with iron. For example, the anti-ulcer drug, cimetidine, which contains an imidazole ring, binds to the iron of cytochrome P_{450}s. This alters the metabolism and hence biological activity of other drugs. Ethanol and a constituent of grapefruit juice, nootkatone, also have an effect on cytochrome P_{450}s. This can lead to an altered pattern of drug metabolism and to drug:drug interactions.

2.9 THE HYDROXYLATION OF AROMATIC RINGS

Many drugs contain an aromatic ring **2.8**. Oxidation of the aromatic ring by the cytochrome P_{450}s leads to the formation of a highly reactive, strained arene oxide **2.9**. In the presence of acid, the epoxide undergoes hydrolysis to a diol **2.10**, which may then be dehydrated to generate the aromatic ring of a phenol **2.12**. Alternatively, a rearrangement may occur with the formation of a ketone **2.11** and thence by enolization, a phenol **2.12**. The rearrangement of a hydrogen atom implicit in this process is known as the NIH shift after the National Institute of Health (USA) where it was discovered. The nucleophile in the cleavage of the epoxide may be the nitrogen of a nucleic acid base or the side chain amino group of a protein. The effect of this new bond is to attach the drug to the nucleic acid **2.13** or protein. This can lead to liver damage (hepatotoxicity). The nucleophile may also be the thiol of glutathione **2.4**. This can lead to deactivation of a toxic metabolite.

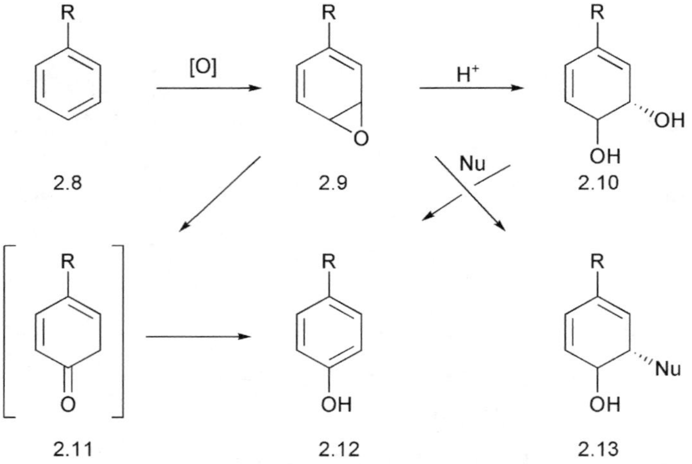

There are many examples of this step. Acetanilide **2.14** is a weak pain-killer (analgesic), but it is converted by the liver to paracetamol **2.15**, a more powerful analgesic. This is an example of a pro-drug:drug relationship. The aromatic hydrocarbon, benzopyrene **2.17** is converted to the carcinogenic 3,4-epoxy-7,8-dihydroxybenzopyrene **2.18**.

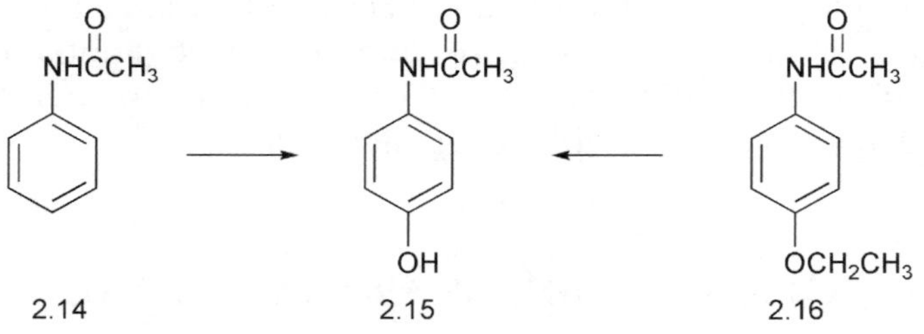

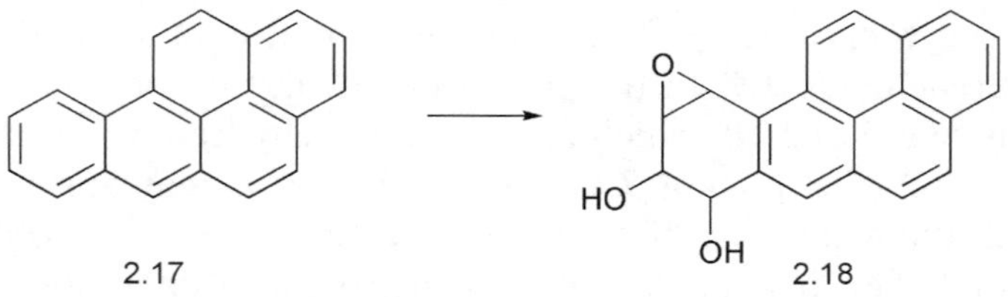

Nucleic acid bases such as guanosine react with the epoxide ring in **2.18**. The aromatic ring of phenobarbital **2.6** may be hydroxylated. The benzylic position adjacent to an aromatic ring is also hydroxylated quite easily. An alkene may be epoxidized by the same process as exemplified by the metabolism of the tricyclic anti-depressant carbamazapine **2.19** to the epoxide **2.20**. A combination of benzylic hydroxylation and epoxidation can be found in the metabolism of safrole **2.21** to the toxic hydroxy epoxide **2.22**.

2.19   2.20

2.21   2.22

2.10 THE HYDROXYLATION OF ALIPHATIC SYSTEMS

The hydroxylation of an aliphatic CH_2 can also be a metabolic step catalysed by these enzyme systems. These oxidations may take place at the end of an alkyl chain (the ω-position) or at the ω-1 position. An example is the hydroxylation of pentobarbital **2.23–2.24**. In another example the ethyl group of the pain-killer phenacetin **2.16** is removed *via* the hemi-ketal, to release paracetamol **2.15**, in a pro-drug:drug relationship.

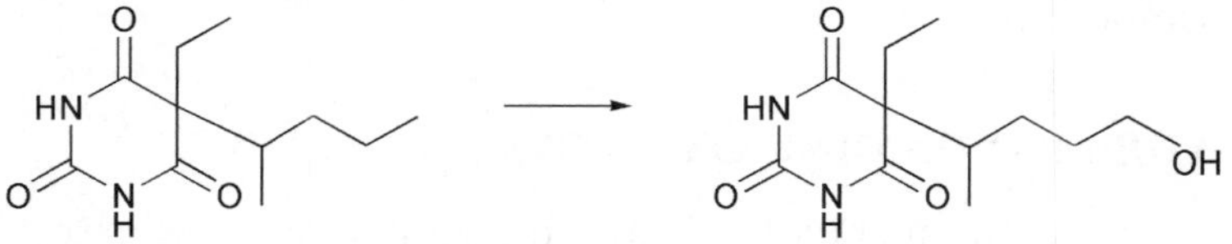

2.23   2.24

These hydroxylations may be the prelude to conjugation. For example, the tranquillizer chlorpromazine **2.25** is hydroxylated **2.26** and then converted to its sulfate ester **2.27**.

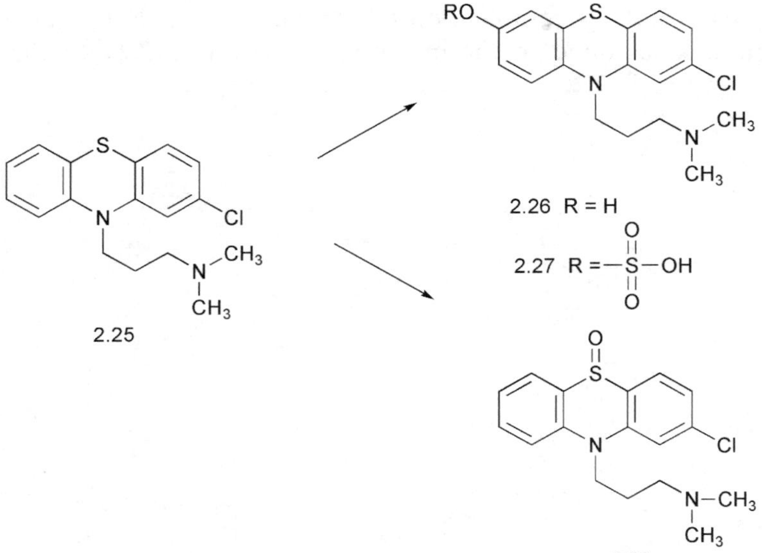

2.26 R = H

2.27 R = $-\overset{\displaystyle O}{\underset{\displaystyle O}{S}}-OH$

2.25

2.28

Other oxidations brought about by these mixed function oxidases involve the conversion of amino and amide groups to amine oxides and hydroxylamines. Thioethers are converted to sulfoxides and sulfones. Examples of the former include the conversion of nicotine **2.29** to its *N*-oxide **2.30** while the thiazine of chlorpromazine **2.25** is converted to a sulfoxide **2.28**. These oxidations can also lead to dealkylation of nitrogen, for example, in the removal of one of the methyl groups from the nitrogen of chlorpromazine **2.25**.

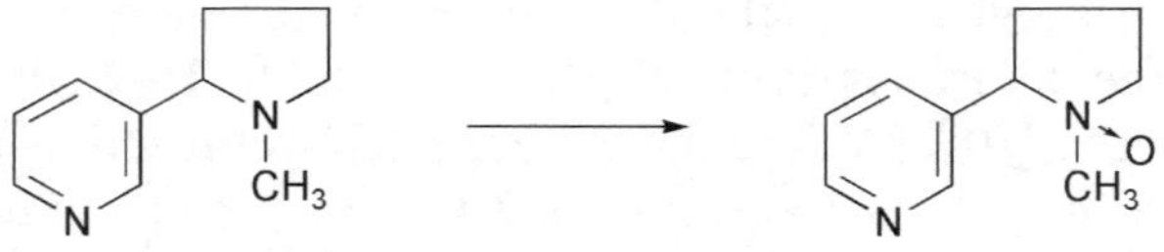

2.29   2.30

The alcohol dehydrogenase is another important oxidative system in drug metabolism.

2.11 THE MONOAMINE OXIDASES

Many amines are metabolized by the monoamine oxidase system (MAO). This enzyme system brings about the dehydrogenation of an amine ($>CH.NH_2$) to an imine ($>C=NH$), which is then hydrolysed to a carbonyl group. The monoamine oxidases are flavin-dependent systems, which play an important role in the metabolism of the neurotransmitters such as dopamine and many of the drugs, which interact with them. There are two broad families of monoamine oxidase. MAO-A for which serotonin is a substrate and MAO-B for which phenyl-ethylamines, *e.g.* dopamine, are substrates. In this case the corresponding aldehydes are formed. The monoamine oxidases are not always plentiful in the liver and it is possible to 'flood' this system, for example, with amine-rich foods.

2.12   OTHER PHASE ONE CHANGES

Although demethylation is a common metabolic process, methylation is also found particularly in the metabolism of the catecholamine neurotransmitters. Catechol *O*-methyl transferase (COMT) is a major metabolic deactivation for the catecholamines and its consequences are discussed later (Chapter 4) in the context of these compounds.

Carboxyl esterases and peptidases lead to the deactivation of many drugs. For example, the ester procaine **2.31**, which is an anaesthetic, is hydrolysed to *p*-aminobenzoic acid **2.32**. The metabolism of the local anaesthetic, lidocaine (xylocaine) (Scheme 5) brings together several of the steps that have been outlined.

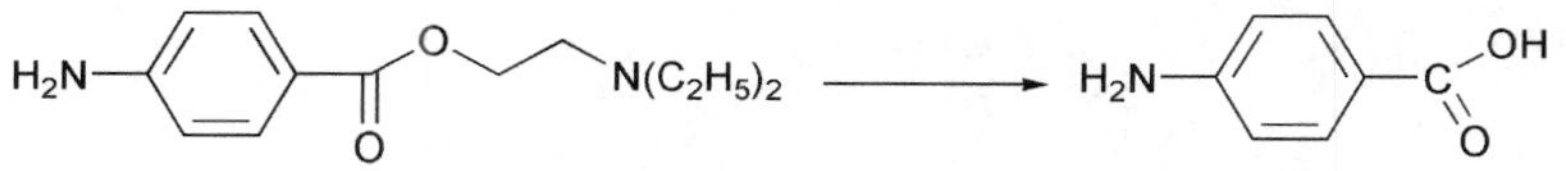

2.31   2.32

Drug metabolism can bring about a change in activity. Iproniazid **2.33** is an anti-depressant while its dealkylation product, isoniazid **2.34** has anti-tubercular activity.

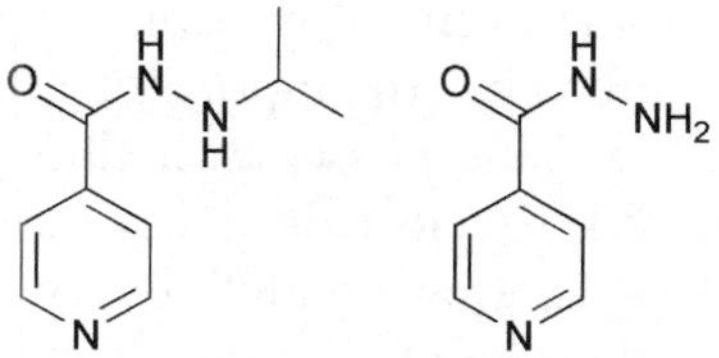

2.33   2.34

Sufficient data on drug metabolism is now available for it to be possible to use molecular modelling techniques to predict the likely metabolism of a drug. Some cytochromes have been cloned and bacterial models can be used to identify putative mammalian metabolites. A drug may be incubated with a microorganism containing the common mammalian cytochromes as a first step in the identification of its metabolites.

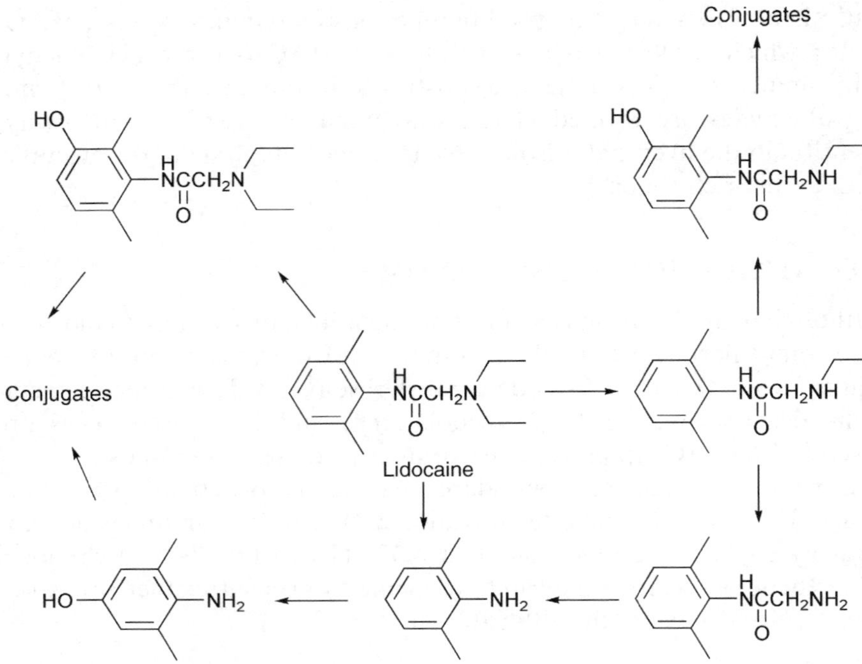

Scheme 5 *The metabolism of lidocaine*

2.13   PHASE TWO CHANGES

The phase two changes involve linking a water-solubilizing group to the drug. The common groups are glucuronic acid **2.2**, which is provided by uridine-5-diphosphate α-D-glucuronic acid, the sulfate **2.3**, which is provided by 3'-phosphoadenosine-5'-phosphonosulfate and the amino acids, glycine **2.1** and taurine. An important conjugation utilizes the tripeptide glutathione **2.4**. This possesses a thiol, which is a powerful nucleophile that can add to electron-deficient carbon atoms. Since electron-deficient centres can be created by oxidative processes and might otherwise be attacked by free amino groups of nucleic acids and proteins, the glutathione conjugation serves to protect the liver from damage. An example is provided by the metabolism of paracetamol **2.15**. Oxidation of paracetamol takes place on the nitrogen with the formation of an *N*–OH **2.35** and this leads, after elimination to the quinone-imine **2.36**. The latter is very electron-deficient and readily adds nucleophiles such as the NH_2 from a protein. The consequence of this addition is that the paracetamol becomes bound to the liver causing liver damage. Glutathione **2.4** provides protection against this by adding to the quinone imine **2.36** to give a conjugate **2.37**. If there is insufficient glutathione in the liver and the amount available is exhausted as in a paracetamol overdose, liver damage with potentially fatal consequences, ensues. Treatment involves the rapid administration of a thiol containing drug such as methionine or acetylcysteine.

2.15   2.35   2.36   2.37

2.14 EXCRETION

A drug and its metabolites may be excreted from the liver into the bile and then into the intestine. Alternatively it may be excreted *via* the circulatory system and the kidneys into the urine. A few drugs are eliminated *via* the lungs. Problems may arise from the elimination of a drug *via* the kidneys. Metabolites have been known to crystallize in the kidneys. The acetate of the anti-bacterial agents, sulfanilamide does this and can produce a blockage in the kidneys.

2.15 PRO-DRUGS

A pro-drug is a pharmacologically inert substance, which is converted in the body to an active drug. Pro-drugs rely on metabolic steps such as those outlined above to release the active drug. Thus a pro-drug is a masked latent drug. This masking of the drug can modify its solubility and alter its distribution allowing it to reach its site of action more effectively. It may enable the compound to cross barriers such as the blood:brain barrier and it may prevent the deactivation and excretion of a drug. It can also produce a slow-release form of the drug.

One of the simplest examples of a pro-drug is that of methenamine (hexamethylenetetramine) **2.38**, which was used as a urinary tract antiseptic because it was broken down in the acidic medium of the urine to release formaldehyde. Another example was the analgesic phenacetin **2.16** in which the ethoxy group was hydroxylated to form a hemi-acetal, which then released paracetamol **2.15**. However phenacetin is no longer available because another of its metabolites has turned out to be toxic. The discovery of the sulfonamide anti-bacterial agents arose from the conversion of the dyestuff prontosil red **2.39** into sulfanilamide **2.40**. Another sulfa drug to build on this is sulfasalazine **2.41**, which was developed in the 1940s to treat ulcerative colitis. The pro-drug was cleaved by amidases in the colon to release sulfapyridine **2.42** and 5-aminosalicylic acid **2.43**. It enabled a useful concentration of sulfapyridine to be established at the site of action.

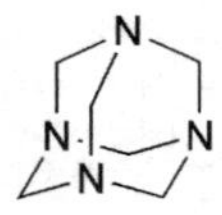

2.38 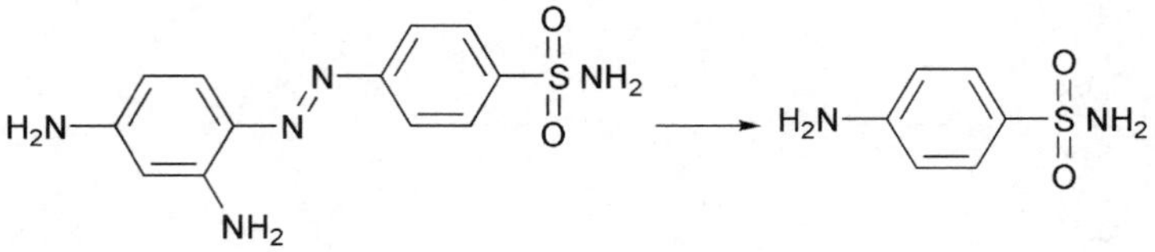   2.39   2.40

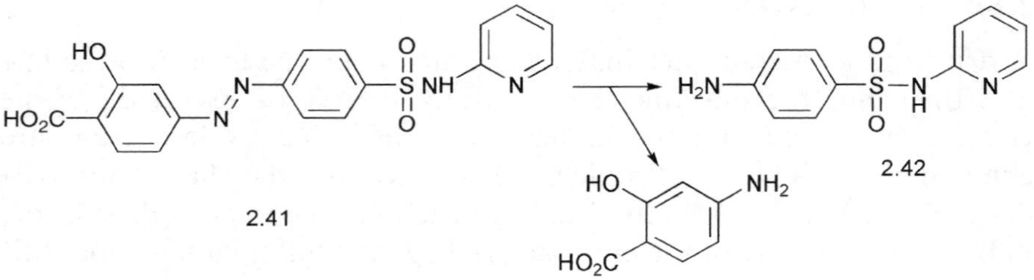

2.41

2.42

2.43

Hydrocortisone **2.44** is used to alleviate inflammation at various sites in the body. The sodium succinate C-21 ester **2.45** provides a water-soluble pro-drug while the 21-palmitate **2.46** is a lipid-soluble pro-drug enabling it to reach different sites of activity. Another example is the anti-viral agent acyclovir **2.47**, which is transported into a virus before it is phosphorylated to give the active species. The triphosphate of acyclovir would not cross the viral cell wall. Azathioprine **2.48** is a pro-drug for 6-mercaptopurine **2.49**, which is used as an immunosuppressant.

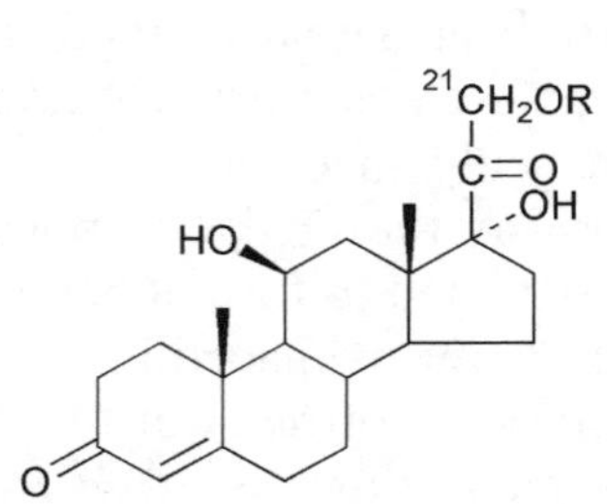

2.44 R = H

2.45 R = $-\overset{\overset{\text{O}}{\|}}{\text{C}}CH_2CH_2$$\overset{\overset{\text{O}}{\|}}{\text{C}}$$-O^-Na^+$

2.46 R = $-\overset{\overset{\text{O}}{\|}}{\text{C}}$(CH$_2$)$_{14}CH_3$

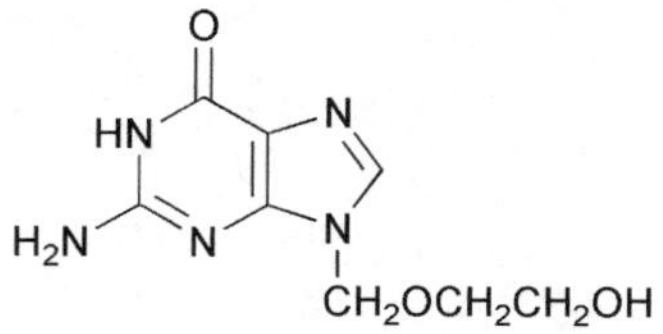

2.47

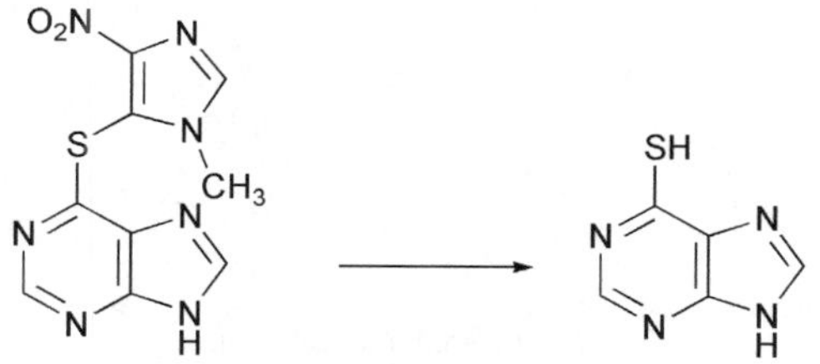

2.48   2.49

2.16 QUANTITATIVE STRUCTURE: ACTIVITY RELATIONSHIPS

Physico-chemical features associated with a structure such as polarity and hydrogen bonding make a considerable contribution to the biological activity. Although there is an overlap, it is possible to discuss these in terms of those factors that affect the transport of a drug to its site of action and those that affect the expression of the biological activity at the site of action. There have been numerous attempts to quantify the relationships between chemical structure and biological activity in terms of measureable physico-chemical parameters and then to use these in a predictive sense in the design of novel drugs.

One of the first attempts to achieve this correlation between biological activity and a measureable physico-chemical parameter was to relate the concentration of various anaesthetics such as diethylether and chloroform required to produce narcosis in test subjects (mice and tadpoles) with the partition co-efficient of the anaesthetic between olive oil and water. This work by Meyer and Overton and by Baum (1899–1901) provided a model for the uptake of the substance by a lipid-like barrier. Many other attempts were made to relate the oil:water partition coefficient to biological activity such as, for example, with a series of barbiturates. The correlation of the oil:water partition co-efficient with biological activity can be a curve with a maximum (Scheme 6). On the one hand too small a partition co-efficient would imply too little material crossing lipid barriers while too high a partition co-efficient

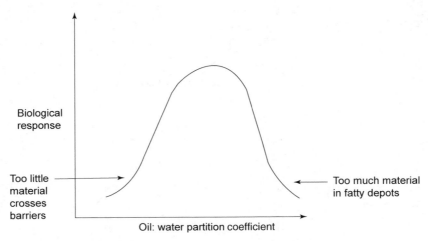

Scheme 6   *The origin of an optimum log P*

would suggest that depots of the drug in fat tissue might be formed. This would lead to insufficient free biologically active material.

Studies on the local anaesthetic action of the alkamine esters of benzoic acid and *p*-aminobenzoic acid **2.32** such as amylocaine and procaine **2.31**, revealed the relationship of both pK, and hence the concentration of free base, and the partition co-efficient between olive oil and water, to the biological activity. The effect of chain branching, a possible steric factor, was also noted. The olive oil:water partition co-efficient measurements were later replaced by measurements based on octanol:water. By the 1960s, the physico-chemical parameters that might be considered in drug design were grouped into three areas, hydrophobic, electronic and steric.

2.17   HANSCH QSAR ANALYSES

A systematic analysis by Hansch of large body of data concerning aromatic compounds led to the establishment of a series of QSAR possessing predictive capability. The work started (1964) by rationalizing the substituent effects on the plant growth promoting properties of phenoxyacetic acids and second, by examining the bacteriocidal action of relatives of chloramphenicol **2.50**. It was then extended to many other series of compounds such as the toxic action of substituted phenols on bacteria and the insecticidal action of substituted benzoic acids. Hansch considered the contributions of substituents initially just in terms of their electronic and lipo-hydrophilic contributions. Steric effects were introduced later.

$$O_2N-\underset{\underset{\text{2.50}}{\underset{|}{OH}}}{\bigcirc}\underset{\underset{CH_2OH}{|}}{\overset{\overset{NHCOCHCl_2}{|}}{}}$$

In physical organic chemistry, the Hammett equation related the rate of a reaction for a substituted aromatic compound to the unsubstituted parent in, for example, the hydrolysis of benzoic acid esters, to a combination of a substituent constant (σ) and a reaction constant (ρ).

$$\log (k_x) - \log (k_H) = \sigma_x \rho$$

The substituent constant (σ_x) can be considered to be a combination of the field and resonance effects. This was used to relate the strengths of aromatic acids and bases.

$$\text{benzoic acids } pK_a = 4.20 - 1.00\sigma$$
$$\text{anilines } pK_a = 4.58 - 2.90\sigma$$
$$\text{phenols } pK_a = 9.92 - 2.33\sigma.$$

Hansch defined a substituent constant π to reflect the effect of a substituent on the lipo-hydrophilic character of the molecule where $\pi_x = \log(P_x) - \log(P_H)$ in which P_x and P_H are the octanol:water partition coefficients for the substituted and unsubstituted compounds. He then produced equations relating the molar concentration, C, of the compound required to produce the biological effect to π and σ which have been expressed in the form: $\log(1/C) = a\pi + \rho\sigma + c$ where ρ in this case relates to the bioassay.

Although these ideas were initially developed in the context of aromatic substances, the inclusion of steric terms allowed the reasoning to be extended to aliphatic systems. Various measurements of steric terms have been introduced such as the molecular refractivity. It is defined as

$$MR = \frac{n^2 - 1 \, \text{mol.wt.}}{n^2 + 2d}$$

where n is the refractive index and d is the density of the compound. This provides a measure of the polarisibility of a group. Another measure of steric factors is known as the Taft steric parameter (E_s) and is based on the effect of a substituent on the rate of acid-catalysed hydrolysis of the substituted methyl acetate.

$$E_x = \log(kXCH_2CO_2CH_3) - \log k(CH_3CO_2CH_3)$$

2.18   CRAIG PLOTS AND THE TOPLISS DECISION TREE

Various schemes have been developed to exploit the Hansch relationships. In the first, known as the Craig plots, π is plotted against σ for a series of different substituents (see Scheme 7). This has the effect of highlighting substituents with potentially similar effects on the Hansch equation. The second scheme is known as the Topliss decision tree. It is based on using the change in biological activity arising from inserting one substituent to suggest the next substituent in order to maximize the chances of synthesizing the most potent compounds in a series as early as possible. For example, in many systems activity increases with increasing lipophilicity, *i.e.* they are $+\pi$ dependent. Hence an initial step in examining a series of aromatic compounds might be to insert a chloro substituent para to a side chain in place of a hydrogen. If this does indeed have a positive effect on the biological activity, it suggests that the potency may be related to a $+\pi$ or $+\sigma$ effect. The biological activity might then be enhanced by preparing the 3,4-dichloro analogue and perhaps the 3-CF_3,4-Cl and 3-CF_3,4-NO_2 analogues. If, however, the introduction of the chlorine led to an analogue that was only equipotent with the parent compound, then this might be interpreted as a $+\pi$, $-\sigma$ effect. The scheme then suggests a series of 4-methyl analogues that possess $+\pi$,$-\sigma$ values. When the 4-chloro analogue is less active than the unsubstituted compound, it may be that there are unfavourable steric effects or that the activity is controlled by $-\sigma$ or $-\pi$ effect. The introduction of a 4-OCH_3 substituent with a $-\sigma$ effect would then be a possible analogue to explore.

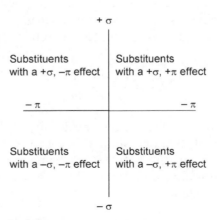

Scheme 7   *Craig plot of π against σ*

2.19 DRUG: RECEPTOR INTERACTIONS

Although the Hansch reasoning does apply to the interaction of a drug with its receptor, many of the factors that are considered relate to the substance crossing barriers and entering different cellular phases. When we consider a drug interacting with a receptor or an enzyme system, there are a number of physico-chemical features that must be taken into account. A receptor or an enzyme system comprises a highly structured chiral environment dominated not just by the pockets created by the peptide chain but also by the varying nature of the amino acid side chains. The structure is also determined by the chemistry of any co-enzyme that is also present and by the presence of water molecules that are hydrogen-bonded within the lattice and may need to be displaced as the drug binds. It is now possible to obtain X-ray structures of drugs bound within the active site of an enzyme and hence identify crucial interactions. Similar information can also be obtained from a nuclear magnetic resonance (NMR) spectroscopy.

Hydrogen bonding interactions with serine and tyrosine residues and ionic interactions with glutamate or aspartate residues are clearly important in binding. However, there are more subtle interactions involving hydrophobic parts of the drug fitting into hydrophobic pockets of a receptor or enzyme, which are created by the alkyl side chains of leucine and valine residues. Some of these pockets may not be significant when the natural substrate binds but may provide additional binding sites for a drug. There are also interactions between aromatic rings in the drug and the aromatic amino acids. The chemistry of interactions involved in the enzyme mechanism and the role of co-factors have to be considered. The interaction of a drug may be competitive with the natural substrate. In such a situation the drug will have many of the binding features of the natural substrate but lack an operational feature. In some circumstances it may be modelled on the transition state for the reaction or in other circumstances the drug may be a substrate for the enzyme but be converted into a product that then reacts with the enzyme, a 'suicide' inhibitor.

When the drug binds to its target, it may bring about a change in the conformation of the receptor or enzyme and thus exert its biological effect. The binding may be adjacent to the active site and the modification to the shape of the enzyme, an allosteric effect, may then affect the binding of the natural substrate. The target is a dynamic flexible system and the binding of the natural substrate may require or produce a change in shape. The allosteric binding of a drug may introduce a rigidity, which can alter this flexibility. The action of the benzodiazepines on the GABA receptor (Chapter 4) may be of this type.

When a drug is modelled on a hormone or a neurotransmitter, there is usually a need to develop selectivity so that the drug only interacts with specific receptors and not with all the receptors that are sensitive to a particular neurotransmitter. The natural neurotransmitters possess at least two distinct binding sites, for example, the catechol and the amino group of dopamine. These are separated by a flexible chain. Hence, it is possible to envisage these binding in different conformations in different receptors. Selectivity might then be achieved by imposing rigidity on the drug structure by inserting additional rings or substituents.

While the binding may involve hydrogen-bonding, receptors may be differentiated by the presence of additional hydrophobic pockets. Use may be made of this by introducing hydrophobic substituents into the drugs to develop selectivity.

This general discussion of the distribution and binding of drugs has highlighted the balance between physical properties that can lead to a successful drug. An interesting analysis of the consequences of this was made by Lipinski who profiled the physical properties of over 2000 successful drugs. After discounting polymers, peptides and phosphates, Lipinski found that approximately 90% of the remaining compounds had a molecular weight less than 500, a calculated log P less than 5, a sum of hydrogen bond donors (OH and NH) less than 5 and a sum of hydrogen bond acceptors (as a sum of N and O) of less than 10. Analysis of the progress of drugs through the clinical phases has shown that there is a convergence towards these figures.

CHAPTER 3

Neurotransmitters as Targets

3.1 AIMS

The aim of this and the following two chapters is to describe the medicinal chemistry of non-infectious diseases that are associated with neurotransmitter and hormonal action. This chapter is concerned with some aspects of the medicinal chemistry of the cholinergic and adrenergic systems. By the end of this chapter you should be aware of:

- the structure of the major neurotransmitters;
- the role of cell-surface receptors in cell signalling;
- the design of selective drugs for specific receptors;
- the structures of some drugs that interact with acetylcholine including neuromuscular blocking agents and local anaesthetics; and
- the structures of some drugs that target adrenergic receptors including anti-hypertensive agents, anti-asthma agents and β-blockers.

3.2 INTRODUCTION

Many actions of the body are at some stage regulated by chemical messengers known as hormones. The receptors for these compounds, which are involved in cell regulation, provide the targets for many drugs. The hormones may exert their biological effect by binding to a receptor or by activating an enzyme system. This receptor may be on the surface of a cell or it may be in the nucleus within the cell. Binding to the cell-surface receptor may initiate a sequence of events either through opening an ion-channel into the cell or by producing a change in shape in a trans-membrane protein that leads to the formation of a second chemical messenger within the cell.

Hormones may be divided into a number of groups. First, there are the circulatory hormones that are produced in one organ such as the

41

hypothalamus or the pituitary gland in the brain or the thyroid or adrenal glands. These hormones then target other organs such as the ovaries, testes, kidneys or the heart. The steroids, adrenalin (epinephrine) and some peptides fall into this class. The endocrine system carries information to target organs by varying the levels of these circulatory hormones in the blood. Second, there are the local hormones whose effects, for example, on inflammation, are at sites adjacent to the cells in which they are produced. Histamine and the prostaglandins fall into this class. These local hormones are sometimes known as autacoids. Many of the situations in which these local hormones are released arise as a response to an event such as a wound. Third, there are the neurotransmitters. These chemical messengers are produced at nerve endings and traverse the very narrow synaptic cleft at the nerve ending targeting the synaptic receptors to bring about the biological effect of a nerve impulse (see Scheme 1). Their role is to mediate the transmission of the information carried by the nerve across the synaptic cleft to the receptor and thus to help to bring about the consequences of the nerve impulse (see Scheme 1). A fourth group of hormones is the second messengers such as cyclic adenosine monophosphate (cAMP), which are produced within the cell as a consequence of the transduction of information across the cell wall in the cell-signalling process. Their role can involve the amplification of the signal of a primary hormone. In addition, there are growth factors and cytokines, which are compounds of higher molecular weight and which affect other cellular processes.

3.3 THE NERVOUS SYSTEM

The nervous system has two major divisions: the central nervous system (CNS) comprising the brain and spinal cord, and the peripheral nervous

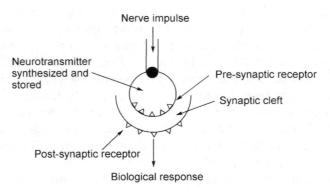

Scheme 1 *Schematic diagram of a nerve ending*

system that lies outside this (see Scheme 2). The peripheral nervous system can be divided into the motor (functional) and sensory nerves. The motor nerves are further sub-divided into the autonomic and somatic nerves. The autonomic nerves regulate the everyday needs of the body such as heart function and gastrointestinal motility, without the conscious participation of the brain. The somatic or voluntary system regulates functions such as muscular and skeletal movements under the conscious control of the brain (see Scheme 2).

The autonomic system is further sub-divided into the sympathetic nervous system, which has the role of preparing the muscle for action, for example, to increase heart rate, while the parasympathetic system has the role of restoring the system to a steady state – for example, decreasing heart rate.

The autonomic system has two sections, which are joined by ganglia or a grouping of nerve cells. Transmission of information from the pre-ganglionic to the post-ganglionic section is mediated by a neurotransmitter. The post-ganglionic neuron also uses a second neurotransmitter at the nerve ending. A neurotransmitter is released and crosses the narrow synaptic cleft to bind to a receptor. This binding to the receptor then brings about the effect of the nerve impulse. The voluntary or somatic nervous system differs from the autonomic nervous system by taking the information directly from the CNS to skeletal muscle without the intervention of ganglia.

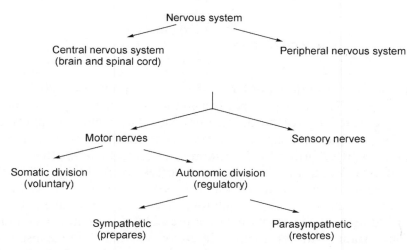

Scheme 2 *Divisions of the nervous system*

3.4   THE NEUROTRANSMITTERS

The major neurotransmitters associated with the autonomic nervous system are acetylcholine **3.1**, norepinephrine **3.2** and epinephrine **3.3** (noradrenalin and adrenalin). Their receptors are known as cholinergic and adrenergic receptors. Whereas acetylcholine is the major ganglionic neurotransmitter, all three neurotransmitters function at nerve endings. Acetylcholine is also the major neurotransmitter for the somatic system.

$$
\begin{array}{c}
\text{O} \\
\parallel \\
\text{CH}_3\text{COCH}_2\text{CH}_2\overset{+}{\text{N}}(\text{CH}_3)_3 \\
\textbf{3.1}
\end{array}
$$

3.2 R = H
3.3 R = CH$_3$

There are other neurotransmitters in the brain and associated with other parts of the nervous system. These include dopamine **3.4**, serotonin **3.5**, γ-aminobutyric acid **3.6** and glutamic acid **3.7**. Histamine, although known primarily as a local hormone, may also have the role of a neurotransmitter. These will form the topic of the next chapters.

3.4

3.5

$$\text{H}_2\text{NCH}_2\text{CH}_2\text{CH}_2\text{CO}_2\text{H}$$

3.6

$$\text{HO}_2\text{C·CH}_2\text{·CH}_2\text{·}\overset{\text{H}}{\underset{\text{NH}_2}{\text{C}}}\text{—CO}_2\text{H}$$

3.7

The neurotransmitters are biosynthesized from amino acids by decarboxylation. Once they have performed their task, they are metabolized and deactivated or reabsorbed (re-uptake). The metabolic pathways may involve the monoamine oxidase pathway or methylation of a phenolic hydroxyl group by, for example, the catechol *O*-methyl transferase (COMT) pathway. In the case of acetylcholine, hydrolysis of the acetyl group by acetylcholine esterase occurs. Another deactivation pathway, which also has a regulatory role, is by re-uptake into the nerve endings or by interaction with a pre-synaptic receptor. Re-uptake is of importance with serotonin (5-hydroxytryptamine) and is discussed in Chapter 4.

3.5 CELL-SURFACE RECEPTORS

The receptors with which the neurotransmitters interact are cell-surface receptors as opposed to the nuclear receptors, which are the target for the steroids. The cell-surface receptors comprise a family of *trans*-membrane proteins. Chemical features involving ionic binding, hydrogen bonding and steric factors can contribute to the way in which a substance binds to these *trans*-membrane protein chains.

The structures of the amino acids that constitute the *trans*-membrane proteins play an important role in creating pockets that are favourable for binding. First, the folding of the protein chain creates a particular steric environment. Second, the hydroxyl groups of a serine or tyrosine or the amide of an asparagine provide hydrogen bonding groups. Third, the carboxyl groups of an aspartate or glutamate provide acidic groups to protonate bases while the basic groups of a lysine or a histidine can bind a carboxylic acid. Some charge transfer can take place to an aromatic amino acid such as phenylalanine while the alkyl side chains of leucine and valine provide hydrophobic interactions.

3.6 ION-CHANNEL-LINKED RECEPTORS

Binding to an ion-channel-linked receptor causes a conformational change in the *trans*-membrane protein, which opens up the ion-channel allowing ions to flow into the cell. The ligand-gated ion-channels comprise two, three or four *trans*-membrane proteins, each with a number of sub-units. The size of the ion-channel and the amino acids that make up the structural protein determine the ions involved. Opening cationic ion-channels involving sodium, potassium or calcium ions leads to excitatory effects. The anionic channels, such as the chloride ion-channel, are inhibitory. Neurotransmitters that bind to ion-channel-linked receptors include acetylcholine, glutamic acid, γ-aminobutyric acid and serotonin. Both acetylcholine and serotonin also bind to G-protein-linked receptors. A number of drugs target these receptors and the ion-channels themselves.

3.7 KINASE-LINKED RECEPTORS

The kinase-linked *trans*-membrane proteins have a single extra-cellular region to which the substrate, often a polypeptide hormone such as growth hormone, may bind. There is a single *trans*-membrane protein chain linked to a tyrosine kinase. This phosphorylation of a tyrosine residue initiates further enzymatic reactions within the cell.

3.8 G-PROTEIN-LINKED RECEPTORS

A large family of receptors are G-protein-linked receptors. These receptors contain seven *trans*-membrane protein loops (see Scheme 3). When a chemical messenger binds to these proteins it can initiate a sequence of events. The change in shape of the *trans*-membrane protein allows a G-protein from within the cell to bind to part of the *trans*-membrane protein within the cell. The G-protein releases a guanine diphosphate and picks up a guanine triphosphate, hence the name G-protein. This change then causes the G-protein to dissociate into three sub-units (α, β and γ). The α sub-units bring about various changes that depend on their structure (see Scheme 4). One sub-unit initiates the formation of cAMP **3.8** from adenosine triphosphate, ATP. Cyclic AMP behaves as a second messenger activating protein kinase A and thence a further sequence of enzymatic reactions (see Scheme 4). Another α sub-unit from a different G-protein activates the enzyme phospholipase C. This enzyme system hydrolyses phosphatidyl inositol diphosphate **3.9** to inositol triphosphate and diacylglycerol (see Scheme 5). The inositol triphosphate mobilizes calcium ions, which in turn affect muscle contraction and cardiac activity. The diacylglycerol activates protein kinase C, which in turn phosphorylates various enzymes involved in the inflammatory response and in tumour propagation.

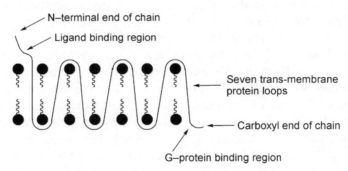

Scheme 3   *Schematic diagram of a G-protein-coupled (linked) receptor*

NH$_2$

3.8

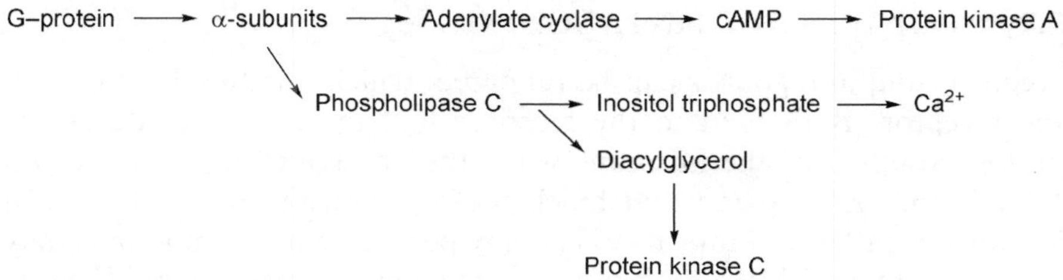

Scheme 4 *Some events in the G-protein cascade*

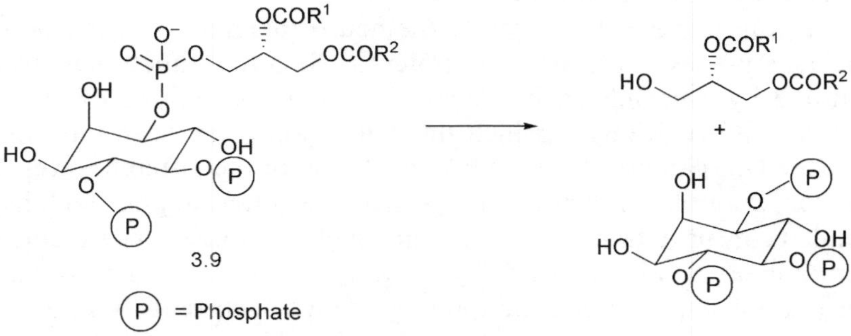

(P) = Phosphate

Scheme 5 *The action of phospholipase C*

The individual ligands bind to different pockets in these G-protein-coupled *trans*-membrane proteins. Many of the adrenergic receptors that respond to noradrenalin (norepinephrine) and adrenalin (epinephrine) and dopaminergic receptors belong to this family. Most of the serotonin receptors and the muscarinic sub-set of the cholinergic receptors also belong to this family.

The role of this chemical intervention in cell signalling is both to amplify the signal from the nerve and also, because the total amount of the neurotransmitter is limited, to regulate the information by preventing an overload.

There are significant differences in the time-scale over which these cell signalling events take place. Effects arising from changes to an ion-channel are very rapid and may occur in milliseconds. Effects involving a *trans*-membrane protein and the initiation of a limited cascade of further enzymatic changes within the cell may occur in fractions of a second to a second. Finally, effects arising from binding of a cell signalling substance to a nuclear receptor and which involve the expression of a nucleic acid fragment in protein synthesis and enzyme action may take minutes to hours to occur. The binding of steroid hormones to nuclear receptors is discussed in Chapter 5.

3.9 AGONISTS AND ANTAGONISTS

Agonists and antagonists can be prepared, which interact directly with the receptor. Both bind to the receptor but an agonist produces the biological effect of the hormone while the antagonist does not. Compounds can be prepared that block the metabolism or re-uptake of a neurotransmitter and thus its effect may persist. Other compounds may interfere either with the biosynthesis or regulatory mechanism. Finally, compounds can be prepared that target the ion-channels themselves or that alter the shape and accessibility of the receptor. Examples of these are discussed in the following sections.

If we consider the structures of the neurotransmitters it can be seen that they possess at least two potential binding functional groups separated by a flexible carbon chain. There are a number of different possible conformations in which the neurotransmitters can bind to the receptor. This can lead to there being sub-sets of the different receptors, one, for example, in which the neurotransmitter might bind in an extended conformation and another in which it is folded. These different receptors may be distinguished by using agonists, which possess a more rigid relationship between the binding groups such that they can bind to one sub-set and not to another. This idea of distinguishing between sub-sets of receptors by using structurally more rigid agonists can be extended to achieving selectivity in drugs that target specific receptors.

3.10 ACETYLCHOLINE

Acetylcholine **3.1** is a neurotransmitter, which is found in both the central and peripheral nervous system. It is biosynthesized from choline and acetyl co-enzyme A by an enzyme choline acetyl transferase. It is stored in the nerve endings and released as required. Acetylcholine is metabolized and deactivated by acetylcholine esterase and the choline may undergo re-uptake for re-synthesis. Neighbouring group participation by the trimethylammonium ion can favour the hydrolysis of the acetate of acetylcholine (see Scheme 6). The inhibition of acetylcholine esterase in the treatment of Alzheimers disease is discussed in Chapter 4.

Two classes of receptors have been distinguished by selective agonists. These are the nicotinic receptors, which are stimulated by nicotine **3.10** and are typically peripheral receptors associated with neuromuscular function. The others are the muscarinic receptors, which are stimulated by muscarine **3.11**. These receptors are found not only in the peripheral system but they also affect heart rate, eye muscle, the respiratory rate and the gastrointestinal tract. There are also important muscarinic receptors in the brain. Examination of the structures of these agonists shows a formal relationship to acetylcholine. In this context it should be borne in mind that the nitrogen atom of the pyrrolidine ring of nicotine readily forms a salt.

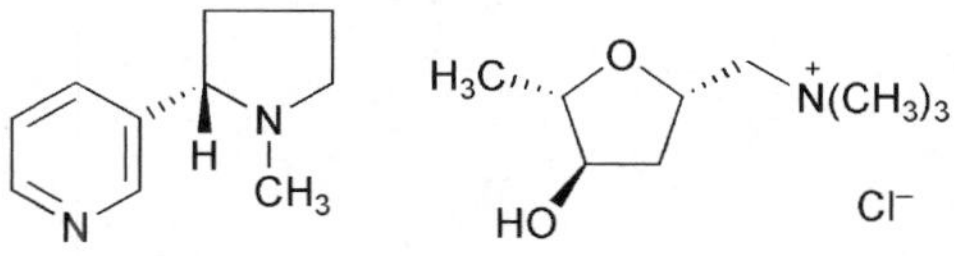

3.10   3.11

3.11   NEUROMUSCULAR BLOCKING AGENTS IN SURGERY

Some of the neuromuscular blocking agents that are used in surgery to relax skeletal muscle act on the nicotinic receptors as cholinergic antagonists. Investigations during the nineteenth century into the mode of action of South American arrow poisons derived from curare led to the identification of the neuromuscular blocking properties of the quaternary ammonium salts of a number of alkaloids, principally tubocurarine **3.12**. These complex alkaloids provided the lead compounds for the development of neuromuscular blocking agents. The properties of quaternary ammonium salts in this context were quite readily investigated with test animals. An important structural feature that was identified was the distance between the two quaternary nitrogen atoms. Because the quaternary ammonium salts are strongly hydrophilic, they do not cross the blood:brain barrier and hence do not affect the CNS where acetylcholine is also involved as a neurotransmitter. Pancuronium bromide **3.13** is an example in which the steroid framework is used to impose some rigidity on the acetylcholine-like moiety.

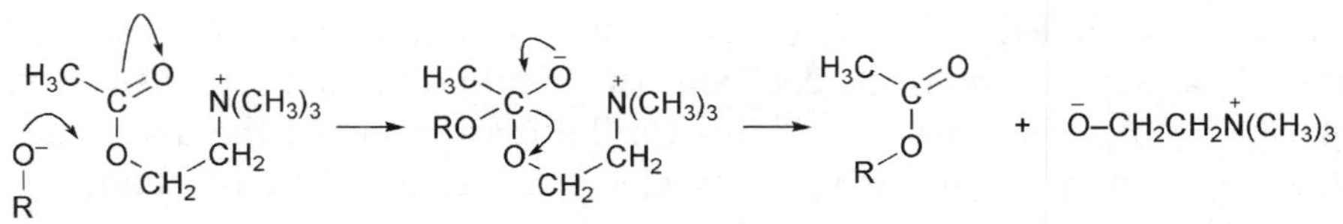

Scheme 6   *Neighbouring group participation in the hydrolysis of acetylcholine*

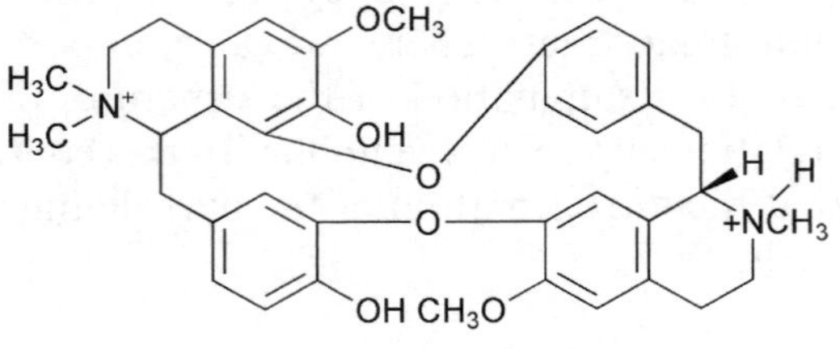

3.12　　2Cl⁻

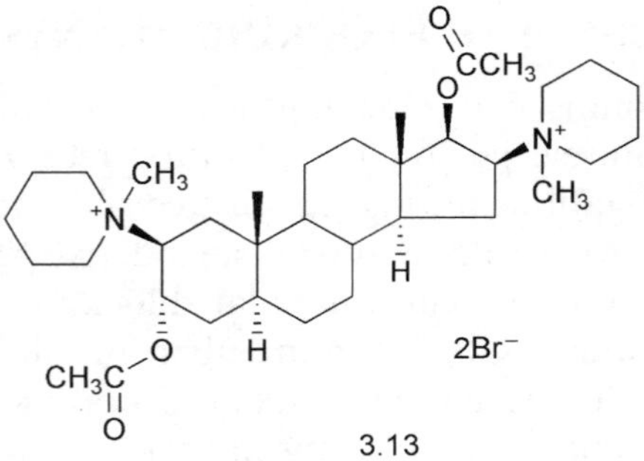

3.13　　2Br⁻

For many uses (*e.g.* in dentistry) a quick acting, short duration muscle relaxant is required. Hence, it was important to design a compound that was not only selective but also readily deactivated. Atracurium benzenesulfonate **3.17** was developed from these studies. Not only is it easily hydrolysed but it also undergoes a rapid Hofmann elimination **3.18** to inactive products. This elimination is facilitated by the carbonyl group of the ester, which renders the adjacent hydrogens acidic. Despite its apparent complexity, it is made by a simple synthesis from pentane-1,5-diol **3.14**, acryloyl chloride **3.15** and the alkaloid papaverine **3.16**. It is worth noting the flexibility of this synthesis in terms of the analogues that might be prepared for structure:activity studies. In particular, the length of the spacer group between the two quaternary nitrogen atoms may be easily varied.

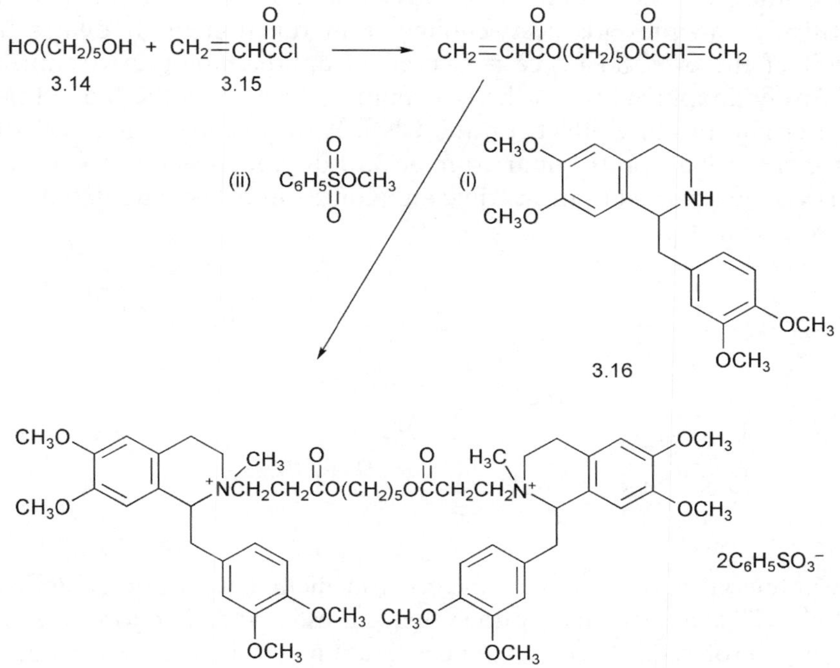

$$HO(CH_2)_5OH + CH_2=CHCCl \longrightarrow CH_2=CHCO(CH_2)_5OCCH=CH_2$$

3.14   3.15

(ii) $C_6H_5SOCH_3$ (i)

3.16

3.17

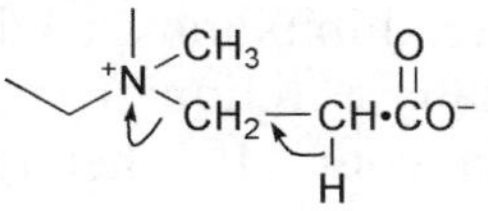

$2C_6H_5SO_3^-$

3.18

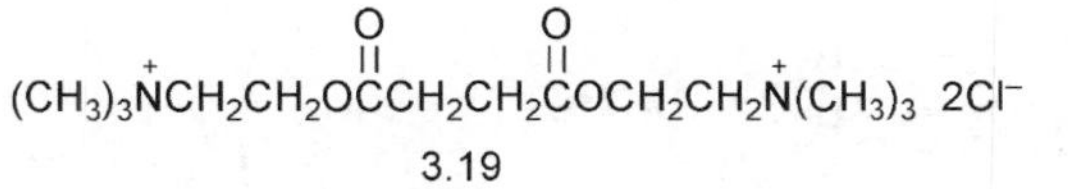

$$(CH_3)_3NCH_2CH_2OCCH_2CH_2COCH_2CH_2N(CH_3)_3 \quad 2Cl^-$$
3.19

A number of local pain-killers and local anaesthetics have acetylcholine fragments. For example, suxamethonium chloride **3.19** is used in a number of throat lozenges.

3.12 MUSCARINIC AGONISTS

Muscarinic agonists are used in ophthalmology in the treatment of glaucoma. The alkaloid atropine **3.20** was obtained from the plant, deadly nightshade, *Atropa belladonna*, which was used in the Middle Ages both because of its properties in dilating the pupils of the eye supposedly to make women more glamorous and also because of its poisonous effects. Atropine is a muscarinic antagonist. It binds competitively and prevents acetylcholine from functioning. It dilates the pupil of the eye and hence it is used in ophthalmology. Atropine is relatively lipophilic and as the free amine, it can cross the blood:brain barrier and produce effects on the CNS. However, quaternary salts of atropine such as ipratropium bromide **3.21** do not cross the blood:brain barrier and are used as muscarinic antagonists in bronchial dilation and muscle relaxation.

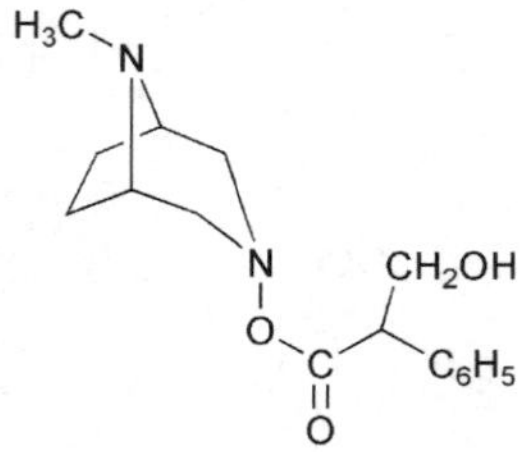

3.20 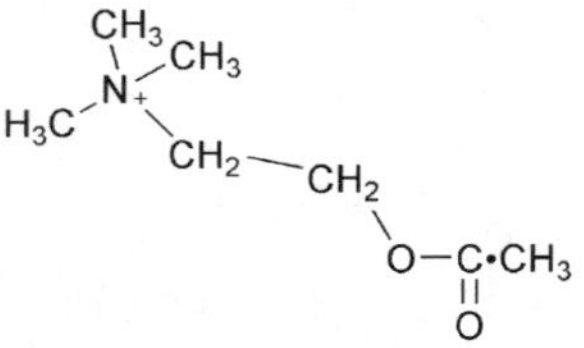 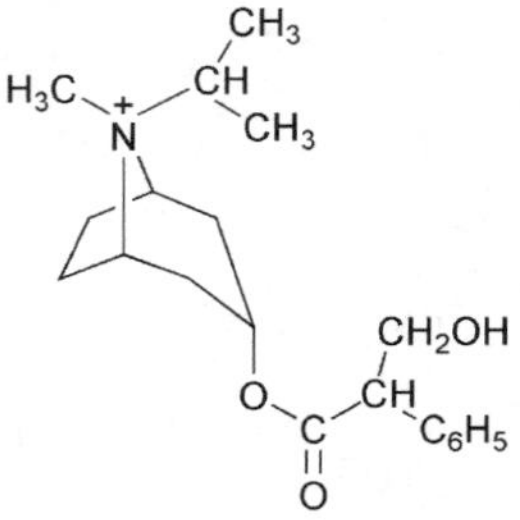   3.21

Cholinergic agonists have some other medicinal applications. Bethanechol **3.22** is a carbamate rather than an ester and therefore it is not easily hydrolysed by acetylcholine esterase increasing its persistence. It stimulates muscarinic receptors and it is used to facilitate urinary expulsion. The alkaloid pilocarpine **3.23** is a γ-lactone and is also stable to acetylcholinesterase. It exhibits muscarinic activity and it is used in ophthalmology to stimulate secretion from the eye and hence reduce intraocular pressure in glaucoma. It is helpful to see a similarity in this structure both to acetylcholine and to muscarine.

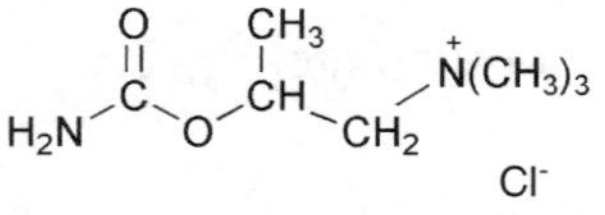

3.22 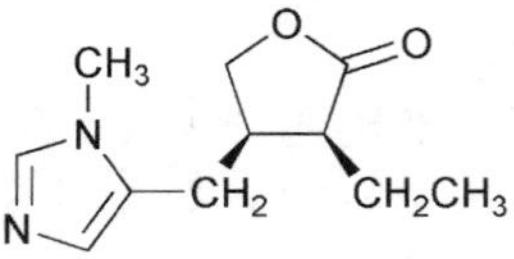   3.23

3.13   LOCAL ANAESTHETICS

Some local anaesthetics have a formal similarity to acetylcholine, although it is possible that they act directly on ion-channels and block sodium ion conductance. The initial leads for these compounds were derived from the anaesthetic action of the alkaloid, cocaine **3.24**, which, despite its addictive properties, became quite a useful drug in the late nineteenth century. Structural simplification led to the development of procaine **3.25** in 1905 and to compounds such as tetracaine **3.26** and benzocaine **3.27**. Another alkaloid isogramine **3.28** was also shown to have local anaesthetic action. Lidocaine (xylocaine) **3.29** introduced in 1948, which is widely used as a local anaesthetic in dentistry and in minor surgery such as the insertion of stitches, was developed from this. Lidocaine and its analogues are easily synthesized. Structure:activity studies revealed the requirement for a lipophilic portion (the aromatic ring) connected by a flexible chain to a hydrophilic unit. X-ray studies revealed the role of the methyl groups in twisting the aromatic ring out of the plane of the amide.

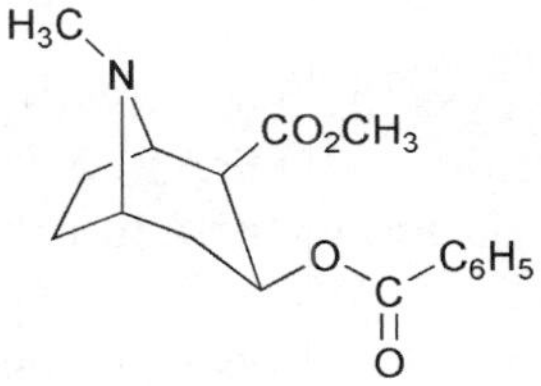

3.24 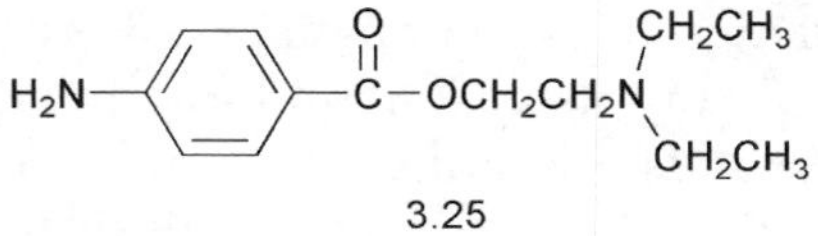   3.25

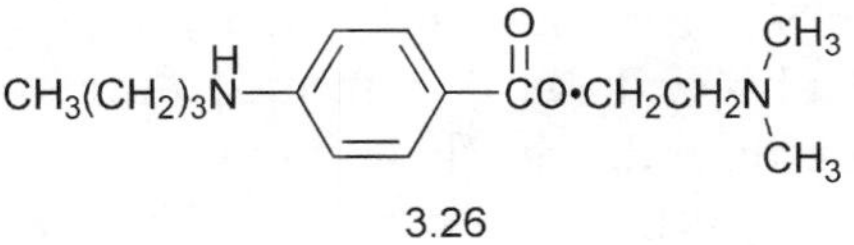

3.26 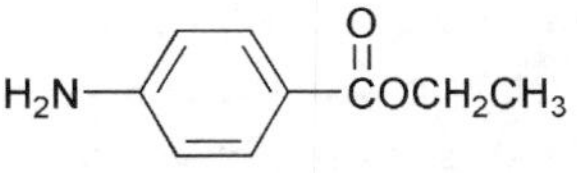   3.27

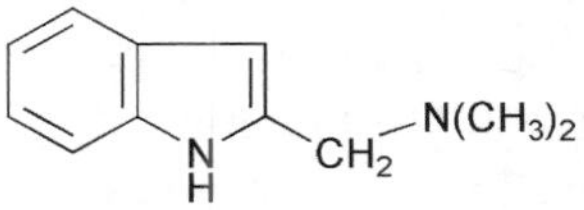

3.28 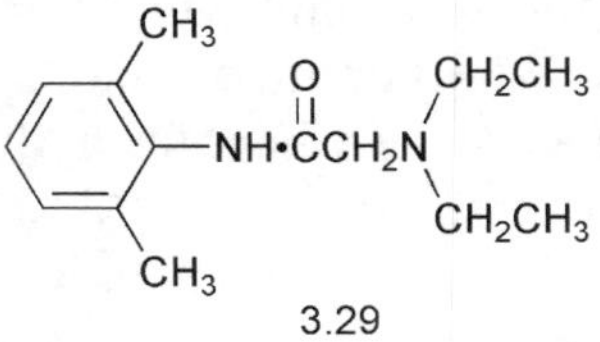   3.29

3.14   CATECHOLAMINES AS NEUROTRANSMITTERS

There are three neurotransmitters that are derived from the amino acid L-tyrosine **3.30** (see Scheme 7). They are dopamine (dihydroxyphenyl-ethylamine) **3.4**, noradrenalin **3.2** and adrenalin **3.3**. Both noradrenalin and adrenalin are chiral alcohols and it is the R enantiomer that is active. They are found in both the peripheral nervous system and the CNS. However, many of the medicinal chemistry studies associated with dopamine involve brain chemistry and will be discussed in the next chapter.

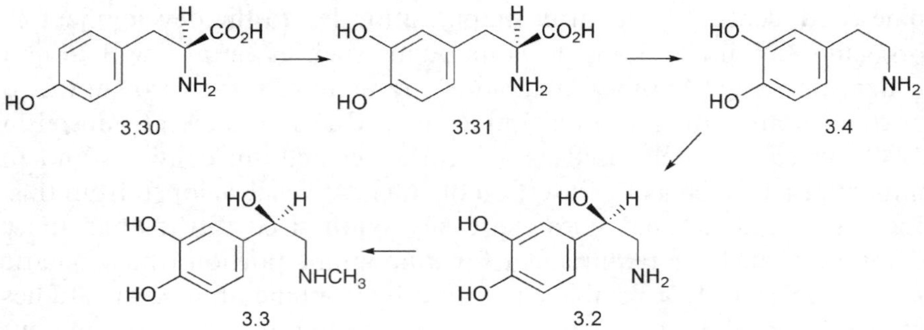

Scheme 7   *The formation of the catecholamines*

The biosynthesis of these compounds involves the hydroxylation of L-tyrosine **3.30** to L-dihydroxyphenylalanine (L-DOPA) **3.31**, decarboxylation to form dopamine **3.4**, hydroxylation to form noradrenalin **3.2** and finally *N*-methylation to give adrenalin **3.3**. The first step in this sequence is rate-limiting. Once the compounds have been formed, they are stored in the nerve endings prior to their release.

While dopamine **3.4** and noradrenalin **3.2** are localized in their action, adrenalin **3.3** is also a circulatory hormone produced by the adrenal glands. The hormones are metabolized by the monoamine oxidase pathway, which converts the compounds via their imines to the aldehydes and which are then oxidized to the acids *e.g.* **3.32**. There are two forms of monoamine oxidase. Monoamine oxidase A metabolizes noradrenalin and serotonin while dopamine is metabolized by monoamine oxidase B. Another metabolic pathway involves the COMT system, which converts the phenolic hydroxyl group that is meta to the side chain, to a methoxyl, *e.g.* **3.33**. The methyl ether may then be metabolized *via* the monoamine oxidase pathway to a hydroxy-acid. The compounds may also be excreted as their sulfates. These metabolic changes prevent the neurotransmitter from binding to its receptor. A regulatory mechanism involves the re-uptake of noradrenalin. This leads to the inhibition of its release and thus it provides a negative feedback regulation.

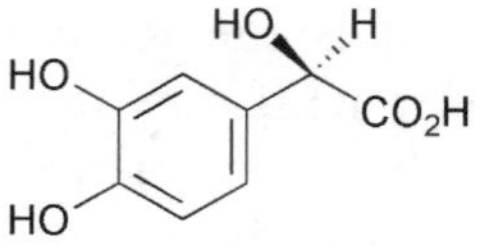

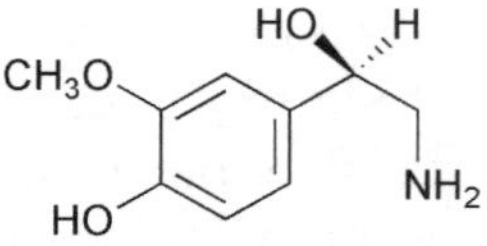

3.15 THE ADRENERGIC RECEPTORS

Although these neurotransmitters are chemically related, they have their own families of receptor, the dopaminergic and adrenergic receptors. In 1948 it was shown by Ahlquist that the adrenergic receptors may be divided by their responses to adrenalin and its analogues into two classes known as the α- and β-receptors. In particular only the β-receptors responded to the analogue, isoprenaline **3.34**. Later, in 1967, Lands was able to show that the β-receptors could be further sub-divided into $β_1$- and $β_2$-receptors. The α-receptors are divided into $α_1$ (post-synaptic) and $α_2$ (pre-synaptic) families.

3.34

When the catecholamine neurotransmitters bind to receptors they produce different responses that depend on the tissue concerned. In general, binding to the α-receptors brings about a constriction while binding to the β-receptors brings about a relaxation of a response. The receptors are G-protein-coupled receptors. The structure of the binding site in the receptor contains some key amino acids. There are two serines whose primary alcohols may act as hydrogen-bond donors to the dihydroxyphenol. The side chain of an aspartic acid unit has a free carboxyl group, which may bind to the amine while there is a phenyl-alanine in which the aromatic ring may interact with the electron-rich aromatic catechol ring. Finally, the β-receptors may be distinguished from the α-receptors by the presence of a hydrophobic pocket into which the isopropyl group of isoprenaline can fit.

Drugs may interact with the catecholamines in various ways, which are illustrated by the following.

3.16 α-ADRENERGIC RECEPTOR AGONISTS

Clonidine **3.35** bears a formal resemblance to β-phenylethylamine. This compound and related imidazolines have attracted interest as anti-hypertensive agents to reduce blood pressure. They act as adrenergic agonists at $α_2$-receptors. Whereas the $α_1$-receptors are post-synaptic receptors, stimulation of which induces the adrenergic pharmacological effects, the $α_2$-receptors are pre-synaptic receptors and binding to them leads to a regulation and an inhibition of the release of the neurotransmitter. Clonidine, while it is an adrenergic agonist, because it is selective for the α_2-receptor it regulates the availability of nor-adrenalin and thus it exerts an anti-hypertensive effect. The chlorine atoms play an important steric role by twisting the imidazoline ring out of the plane of the aromatic ring. Calculations have been made on the optimum conformation of noradrenalin. The distances between the centre of the aromatic ring and the side chain nitrogen are comparable to the distances in clonidine.

An antagonist that has anti-hypertensive activity is labetalol **3.36**. This compound has two asymmetric centres and hence there are four optical isomers. The RR enantiomer is the best for binding to and blocking the β-adrenergic receptor while the S isomer of the benzylic alcohol showed some activity as an antagonist of the α-receptors. These differences reveal the importance of discussing the biological activity of particular enantiomers.

3.35   3.36

3.17 β-ADRENERGIC RECEPTOR AGONISTS – THE DEVELOPMENT OF ANTI-ASTHMA DRUGS

Bronchial asthma is characterized by breathlessness and bronchial con-striction. It may affect as many as 5% of the population. A number of asthma attacks are initiated by the inhalation of an allergen. This induces an inflammatory response in the lungs and a constriction of the bronchial muscle. Treatment involves reducing the response to the allergen, alleviating the inflammatory response and reducing the bron-chial constriction using a bronchodilating agent. Agonists of β-ad-renergic receptors bring about a relaxation of the bronchial muscle but they can also produce an increase in the force and rate of contrac-tion of cardiac muscle. Examination of a series of isoprenaline analogues led to a division of the β-receptors into the β_1 (cardiac) and β_2 (bron-chial) muscle sub-types. Interaction with the β_2-receptor led to a relax-ation of the bronchial muscle. In particular, the t-butyl analogue **3.37** of isoprenaline was considerably more potent on bronchial tissue than on cardiac tissue, *i.e.* it was a more selective β_2 agonist. Another compound, isoetharine with an ethyl substituent on the side chain, was also more active on respiratory smooth muscle than on heart muscle showing that selectivity was possible. However, these compounds were metabolized rapidly by the COMT system. This enzyme system has a substrate specificity for catechols (ortho-dihydroxyphenols). It methylates the phenolic hydroxyl group that is meta to the side chain and so prevents the catechol from binding to the receptor.

Orciprenaline, which has a meta rather than an ortho relationship between the phenolic hydroxyl groups, was a poor substrate for COMT and so it had a longer period of action. A methanesulfonamide group has a comparable acidity to a phenol and therefore might bind to a similar receptor but not necessarily be a substrate for COMT. The methanesulfonamide, soterenol **3.38** proved to be a β-stimulant. What was required was a compound that would still hydrogen bond to the serine units in the receptor but was not a catechol and therefore not a substrate for COMT. In 1966, this led to the synthesis of salbutamol (ventalin®) **3.39** in which a hydroxymethyl group replaced the phenolic hydroxyl. This drug is a selective $β_2$ stimulant with relatively little effect on the heart. It has a longer duration of action compared to isoprenaline. Salbutamol is chiral and it is the R enantiomer that is biologically active.

3.37   3.38

The synthesis of salbutamol **3.39** fulfils several important criteria for the commercial synthesis of a drug. It starts from a cheap starting material, aspirin **3.40**, and it is simple enough to be scaled up. It can be used to synthesize labelled material for metabolic studies and it is sufficiently flexible to be used to prepare analogues for structure:activity studies.

Analogues of salbutamol have been prepared, which have bulkier substituents on the nitrogen. This modification of the structure imparts a greater selectivity for the $β_2$-receptor and increases the lipophilicity of these compounds. An example is salmeterol **3.41**.

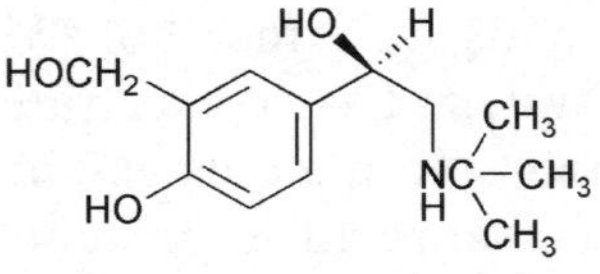

3.39

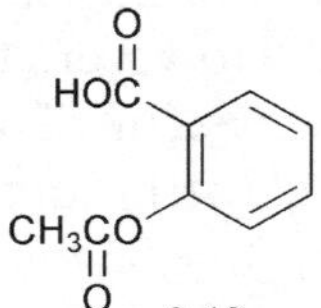

3.40

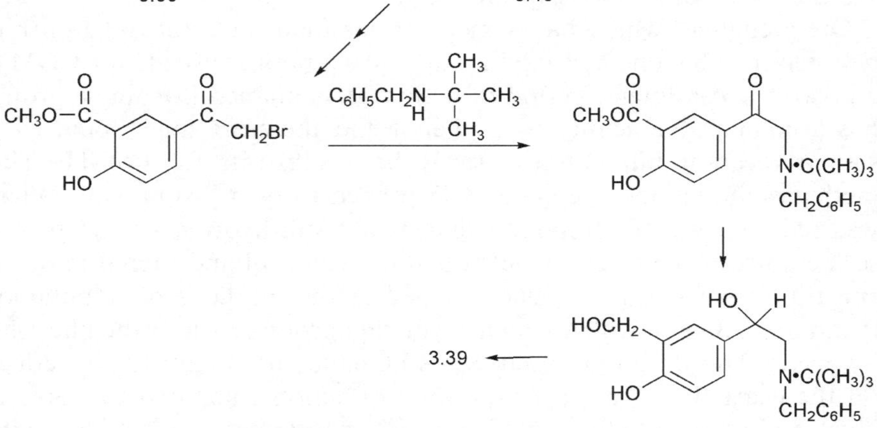

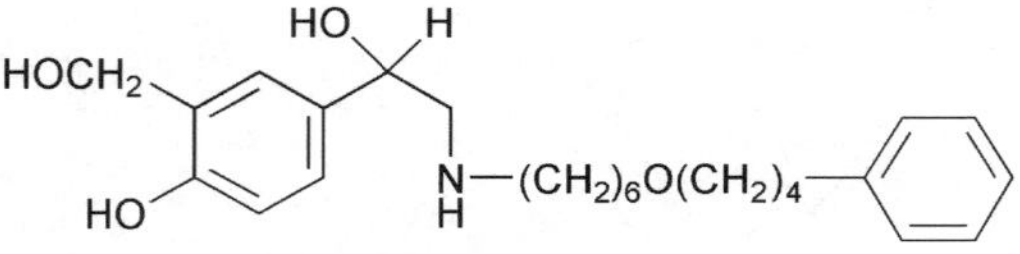

C₆H₅CH₂N—Ċ—CH₃

3.39

3.41

The fact that dihydroxylic phenols with a meta relationship between the hydroxyl groups are not substrates for the COMT has been exploited with terbutaline **3.42** and with fenoterol **3.43**. The latter has a second asymmetric centre in the molecule and in this case the activity resides mainly in the RR isomer.

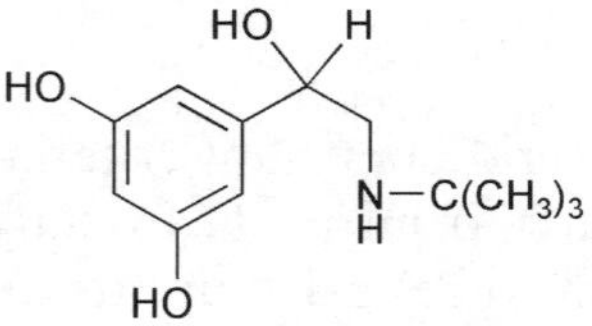

3.42 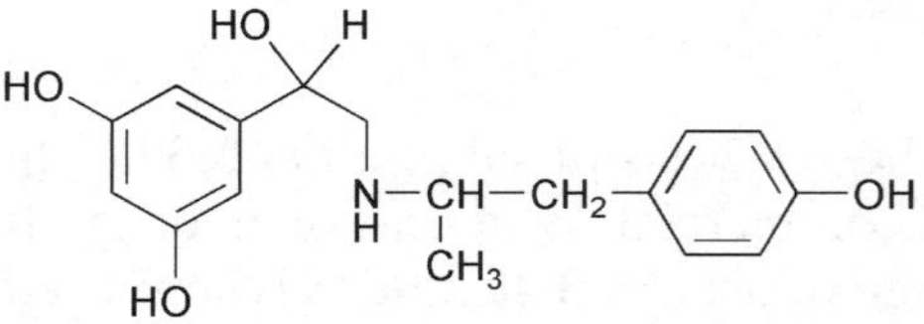   3.43

Some completely different compounds are also used in the treatment of asthma. The discovery of sodium chromoglycate (Intal®) **3.45** was based on the bronchodilating biological activity of khellin **3.44**, a natural product obtained from the Mediterranean plant, *Ammi visnaga*. This bronchodilating agent provides protection against allergen-induced asthma, particularly in children. The anti-inflammatory steroid, fluticasone **3.46**, is used to reduce inflammation in the airways and despite its lack of volatility, it can be administered in an aerosol spray. In this case the action is against a different aspect of asthma, the inflammatory response. Some inhalers, *e.g.* Symbicort® contain a mixture of a corticosteroid such as budesonide **3.47** and a β_2 agonist, such as formoterol **3.48**. An indication of the level of activity of these compounds is that each dose contains 160 µg of the steroid and 4.5 µg of the β_2 agonist. The formoterol utilizes a formamide in place of the phenolic hydroxyl groups. This formamide is not a substrate for COMT.

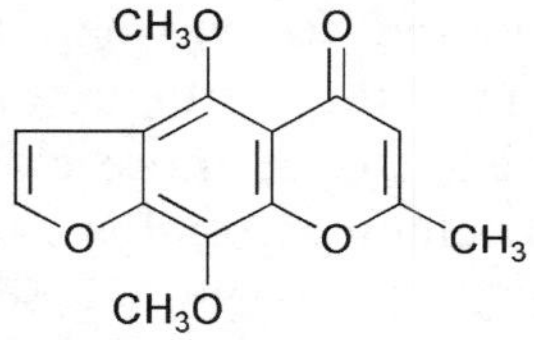

3.44

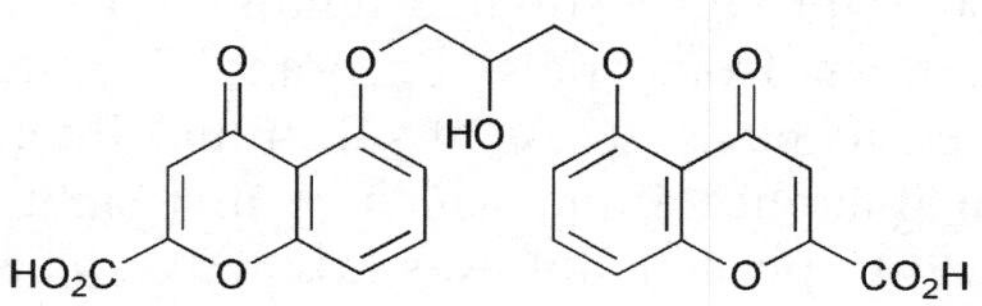

3.45

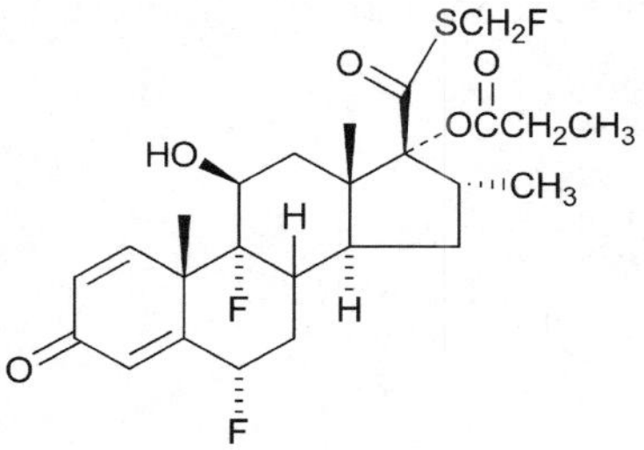

3.46

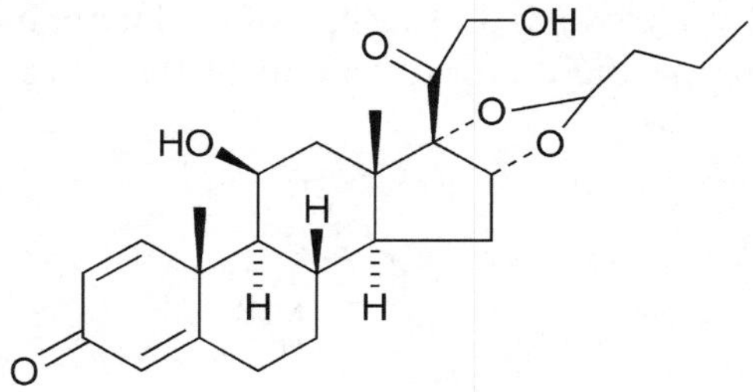

3.47

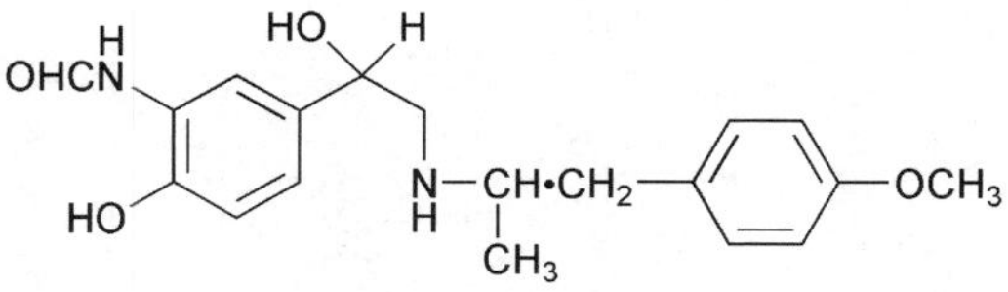

3.48

3.18 β_1-ADRENERGIC ANTAGONISTS 'β-BLOCKERS'

Patients with angina suffer severe chest pain on exercise. Exercise requires the heart to work harder and supply more blood. In order to do this the heart requires more oxygen and nutrients from the blood that is supplying the heart muscle. If the arteries supplying the heart are blocked, there are problems of blood supply. The heart has to work harder to fulfil the need and this overwork eventually produces a heart attack. The cardiac activity producing an increase in heart rate is stimulated by catecholamines acting at the β_1-receptors. A drug, which was an antagonist, a β-blocker, should prevent the heart rate rising by blocking the effect of sympathetic nerve stimulation. The heart's demand for oxygen should then be reduced.

One of the first compounds to show this effect was a partial agonist, dichloroisoprenaline **3.49**, which was prepared in 1958. However, this also had effects on bronchial muscle. In 1961 replacement of the two chlorines by a further aromatic ring gave the naphthalene derivative, pronethalol **3.50**. This compound was a potent β-blocker but unfortunately it produced thymic tumours in mice and in 1963 it had to be withdrawn from trials. Propranolol **3.51** prepared from α-naphthol proved to be a successful substitute. Interestingly, the analogue made from β-naphthol and which at first sight appears to be more closely related to pronethalol, was less active. The side chain alcohol is chiral. Structure:activity studies showed that the activity resided in the R series and that the compounds required a free N–H. Propranolol **3.51** is synthesized from α-naphthol and l-chloro-2,3-epoxypropane followed by cleavage of the epoxide. There has been considerable interest in preparing a chiral version of the drug.

3.49   3.50

3.51   3.52

Propranolol **3.51** is a lipophilic compound and hence it has a high octanol:water partition coefficient. This high lipophilicity is known to facilitate the transfer of compounds across the blood:brain barrier and this could be the reason for some of the side effects associated with the action of propranolol on the CNS. Hence a number of less hydrophobic compounds were synthesized. Sotalol **3.52**, in which the sulfonamide mimics the acidity of a phenol, showed some β-receptor-binding properties. Practolol **3.53**, synthesized from paracetamol, has a log P of 0.71

(corresponding to a partition coefficient of 6:1 between octanol and water) while propranolol **3.51** has a log P of 3.65 corresponding to a partition coefficient of 4500:1. Because it did not cross the blood:brain barrier as easily as propanol, this compound had fewer CNS side effects and was more cardioselective. However, toxicity developed in some patients. This led to the development of atenolol **3.54**, which did not suffer from these problems.

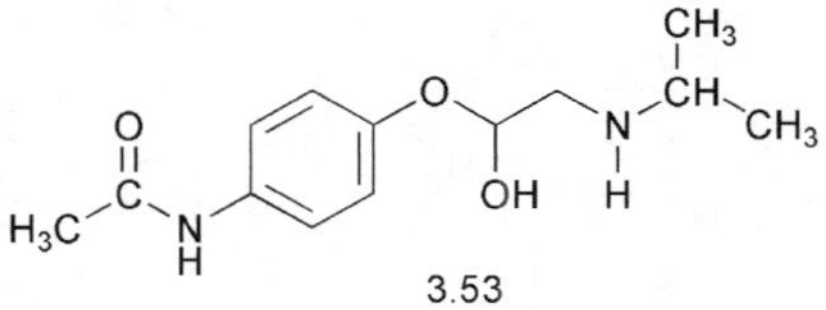

3.53

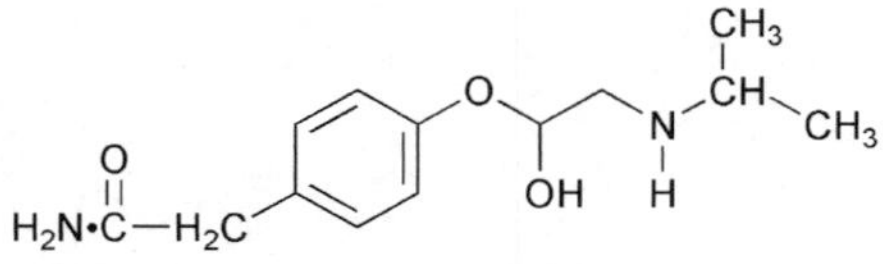

3.54

3.19   THE TREATMENT OF HYPERTENSION

There are other approaches to the treatment of hypertension arising from the heart muscle having to pump too hard. The adrenergic receptors are associated with ligand-gated ion-channels. Opening these channels allows an influx of calcium ions into the cell. This leads to the formation of a complex between the calcium ion and an intracellular protein, calmodulin, which then initiates muscle contraction. Compounds that block the opening of these ion-channels can have a vaso-dilating effect and are then useful in the treatment of hypertension. A number of dihydropyridines introduced in the 1970s such as nifedipine **3.55**, have this action. Although nifedipine can be made by the classical Hantzsch synthesis using 2-nitrobenzaldehyde, ethyl acetoacetate and ammonia, a cleaner product is obtained by a two-step reaction. 2-Nitrobenzaldehyde is condensed with one mole of ethyl acetoacetate to give the benzylidene derivative **3.56**. Addition of the enamine, ethyl 3-aminocrotonate **3.57**, then gives nifedipine **3.55**. Another useful drug, which appears to act as a calcium channel blocking agent is verapamil **3.58**. The more active isomer is the 2S-(-)-verapamil.

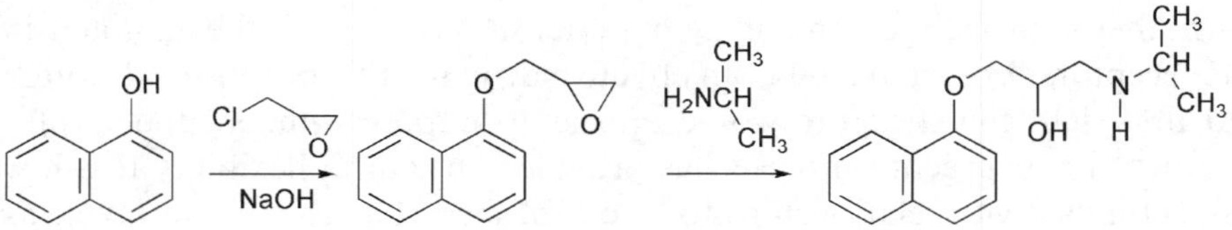

3.51

3.55   3.56   3.57

3.58

High blood pressure, hypertension, can be controlled by β-blockers, by calcium channel blockers or by increasing water excretion with diuretics that act on the kidneys. Compounds that are diuretics can have side effects arising from alteration of the electrolyte balance. A further method is to restrict the formation of the peptide hormone, angiotensin II, which activates receptors located in the smooth muscle of the arterioles that bring about vasoconstriction. Many peptide hormones are formed by the cleavage of longer peptide chains. Angiotensin II is an octapeptide, which is produced as part of the renin–angiotensin cascade (see Scheme 8). Inhibition of the angiotensin-converting enzyme (ACE) by which a decapeptide angiotensin I is converted to angiotensin II provides a means of achieving this. The ACE is a carboxydipeptidase, which cleaves angiotensin I between a phenylalanine and a histidine two residues from the *C*-terminus of the peptide. This enzyme system has a zinc ion at the active site. Captopril **3.59** was designed as a substrate mimic containing an appropriately placed thiol group to bind to the zinc. The synthesis of captopril has to generate the correct absolute stereochemistry. One way in which this was achieved was by the microbiological hydration of methacrylic acid.

The ACE is a relatively non-specific protease and has a number of other substrates and consequently the inhibition can lead to side effects. Another approach to controlling hypertension is to block the angiotensin II receptors. Losarten **3.60** is an effective drug for this purpose. Although at first sight the structure would appear to impose some synthetic difficulties, a convergent synthesis has provided structural flexibility to allow structure:activity relationships to be established. The use of a tetrazole as an acid mimic to impart solubility is an interesting structural feature.

Angiotensinogen

⟵══════ Renin inhibitors

Angiotensin–I

⟵══════ Angiotensin converting enzyme inhibitors

Angiotensin–II

⟵══════ Angiotensin–II antagonists

Vasoconstriction - increased blood pressure

Scheme 8 *The angiotensin–renin cascade*

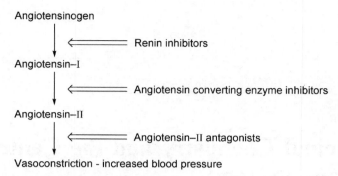

3.59   3.60

CHAPTER 4

Medicinal Chemistry and the Central Nervous System

4.1 AIMS

The aim of this chapter is to describe the role of medicinal chemistry in the alleviation of mental disease. By the end of this chapter, you should be aware of

- the different strategies that are used in correcting for the deficiency of a neurotransmitter in neurodegenerative diseases such as Alzheimer's and Parkinson's diseases;
- the medicinal chemistry used in the treatment of mental diseases arising from an overactive dopaminergic system;
- methods for the treatment of depression based upon the role of serotonin;
- the role of γ-aminobutyric acid in the brain;
- the way in which the benzodiazepines (BZDs) act in the treatment of anxiety; and
- the role of morphine in the control of pain.

4.2 INTRODUCTION

The brain is a site of intense chemical activity. The use of chemical substances to treat the response of the brain to situations, such as pain, to modify the emotions and to treat mental disease has been a continuing feature of medicinal chemistry. The development of drugs to treat mental disease has been one of the major successes of medicinal chemistry in the second half of the twentieth century. Some mental illnesses, such as Alzheimer's and Parkinson's disease, are progressive neurodegenerative diseases producing in the case of the latter a deficiency of the neurotransmitter, dopamine **4.1**. Other diseases involve impairment in the formation or metabolism of a neurotransmitter or a disturbance in the balance between two neurotransmitters in the brain. It is becoming increasingly apparent that a number of conditions such as depression and anxiety involve a disturbance to the balance between several neurotransmitters such as noradrenalin **4.2** and serotonin **4.3**. Furthermore, many of the neurotransmitters have a series of different receptor subtypes and the disturbance may be associated with only one of them. Selectivity of action of medicines becomes very important.

| 4.1 | 4.2 | 4.3 |

Drugs that affect mental processes can be divided into those which are neuroleptic agents, anti-depressants and mood-stabilizing agents and anxiolytic agents. The neuroleptic agents include the major tranquillizers, which are used in the treatment of serious psychoses such as schizophrenia. Many of these act at dopamine receptors in the brain. Some anti-depressants and mood-stabilizing agents act by increasing noradrenalin and serotonin levels in the brain. They achieve this by blocking the metabolism of the neurotransmitters by monoamine oxidase and by affecting the re-uptake mechanisms. On the other hand, a number of anxiolytic agents and other minor tranquillizers such as the BZDs achieve their effect by targeting the γ-aminobutyric acid (GABA) **4.4** receptors.

| 4.4 | 4.5 |

Some neurotransmitters such as glutamic acid **4.5** are excitatory, while others such as GABA **4.4** are inhibitory. A balance between these is needed in the brain. Epilepsy involves an impairment of the GABA system. The relief of pain, analgesia, is associated with several receptors in the brain particularly the opioid and cannabinoid receptors. Compounds, which target these receptors may also bring about a sense of euphoria and have consequently become addictive. In this area, achieving a separation of analgesia and addiction is an important target.

A major barrier in the circulatory system is known as the blood:brain barrier and it prevents some compounds from reaching the brain. This barrier has lipophilic character. Designing compounds to successfully traverse this barrier is an important aspect of the medicinal chemistry of the brain. The presence of this barrier and the complexity of brain biochemistry makes the bioassay of compounds particularly difficult. Animal models still have a role to play.

Not all drugs that affect mental disease target the neuroreceptors. Some act directly on ion-channels and on the ionic balance in cells and the brain. Thus, lithium carbonate is used as a sedative in the treatment of patients with acute psychotic excitement associated with mania.

4.3   THE TREATMENT OF NEURODEGENERATIVE DISEASES

4.3.1   Alzheimer's Disease

Alzheimer's disease is a degenerative neurological disease in which the patients increasingly lose their memory and become confused. It is the most common cause of dementia in the elderly. Nerve cells in the brain particularly in the hippocampus and the cerebral cortex, die and the levels of several neurotransmitters including acetylcholine **4.6**, fall.

The loss of the nerve cells is accompanied by the formation of β-amyloid plaques and neurofibrillary tangles. The plaques are formed from clumps of β-amyloid protein that accumulate in the space between the nerve cells, while the tangles are found inside the nerve cells and are derived from a protein known as tau.

4.6   4.7   4.8

The acetylcholine deficiency cannot be remedied by just giving acetylcholine because it is too polar to cross the blood: brain barrier. Many of the current strategies are aimed at alleviating the symptoms by inhibiting the metabolism of acetylcholine in the brain. Acetylcholine is metabolized by hydrolysis of the acetate ester. Acetylcholine esterase inhibitors that are of value in this connection are tacrine (Cognex®) **4.7**, the daffodil alkaloid galanthamine (Reminyl®) **4.8**, donepezil hydrochloride (Aricept®) **4.9** and rivastigmine (Exelon®) **4.10**. More recent targets have been to prevent the β-amyloid protein undergoing cleavage in a way that initiates the formation of the β-amyloid plaques. This cleavage is brought about by a protease known as β-secretase and a number of compounds have been identified, which inhibit the production of this enzyme. These include the spiroketal antibiotic, monensin.

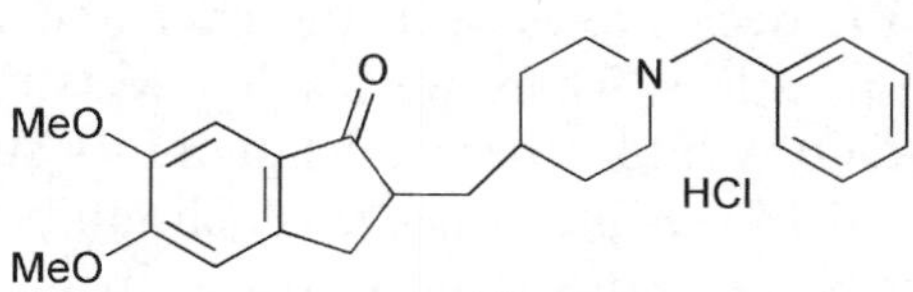

4.9

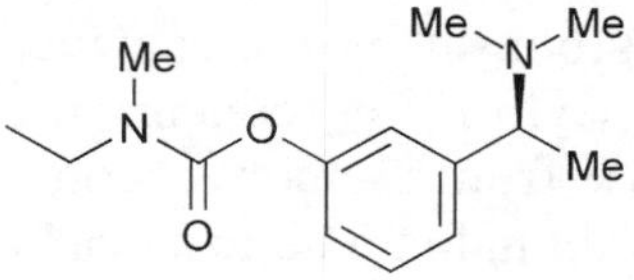

4.10

4.3.2 Parkinson's Disease

Dopamine is a major neurotransmitter in the brain. A dopamine deficiency is associated with Parkinson's disease, while overactive dopaminergic systems are associated with schizophrenia and other mental diseases.

The treatment of Parkinson's disease illustrates several strategies in medicinal chemistry. The disease is characterized by a shaking of the limbs and a typical gait and it has its origins in the loss of cells that produce dopamine and noradrenalin in the regions of the brain known as the substantia nigra and locus ceruleus. Dopamine **4.1** is formed from L-tyrosine **4.11** by hydroxylation to give L-dihydroxyphenylalanine (L-DOPA) **4.12** and decarboxylation. It is metabolized to homovanillic acid **4.13** by the catechol *O*-methyl transferase system *via* **4.14** and by monoamine oxidase B. This knowledge forms the basis of various treatments.

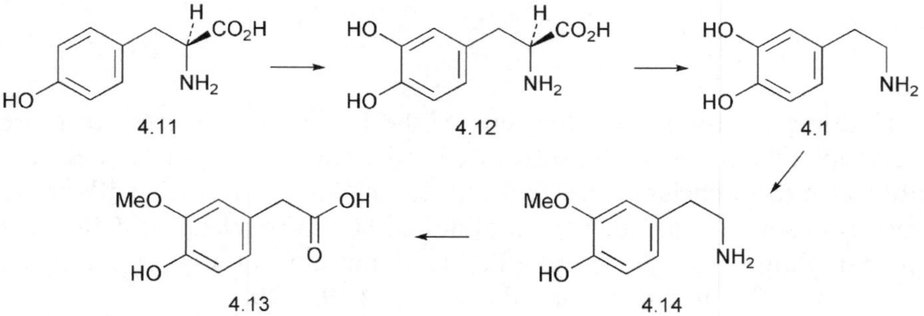

4.11   4.12   4.1

4.13   4.14

The identification, in the late 1950s, of a dopamine deficiency in Parkinson's patients was a major development. However, attempts to use dopamine directly failed not only because it was metabolized too quickly but also because it does not cross the blood:brain barrier. But, success was achieved in 1967 by the administration of the biosynthetic precursor of dopamine, dihydroxyphenylalanine **4.12**. In 1970, the L-stereoisomer L-DOPA, was introduced into therapy. Since it was necessary to administer large doses, attention was directed at increasing the proportion of the L-DOPA that reached the brain by inhibiting its metabolism by the enzyme, DOPA decarboxylase in the liver and the kidney, *i.e.*, by reducing the first-pass loss. There were side effects that arose from the conversion of L-DOPA into other catecholamines such as adrenalin. The drug carbidopa **4.15** was introduced, which inhibited DOPA-decarboxylase. However, because this drug did not cross the blood:brain barrier, dopamine formation still took place in the brain.

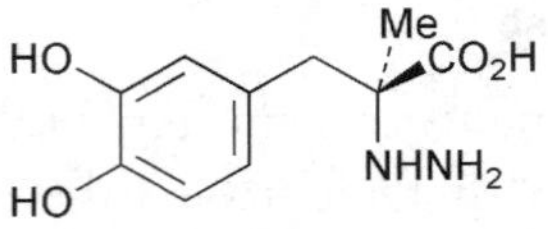

4.15

Inhibition of the metabolism of dopamine also provided a strategy for retaining dopamine in the brain. The monoamine oxidase enzyme system utilizes flavin adenine dinucleotide (FAD) as a co-enzyme. The FAD **4.16** accepts the hydrogen atoms from the substrate and is reduced to FADH$_2$ **4.17**.

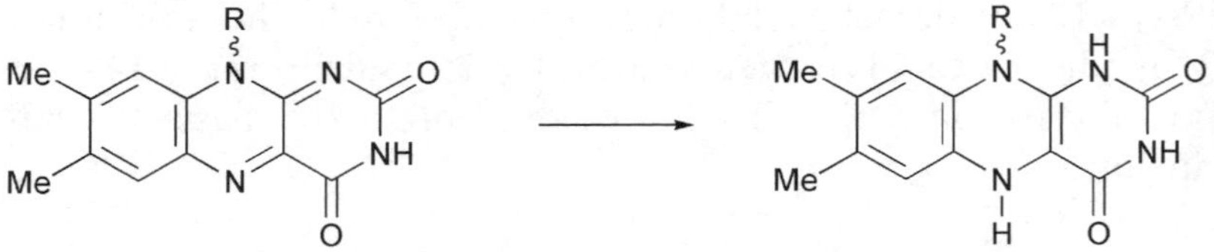

FAD
4.16

FADH$_2$
4.17

In doing so, the imine nitrogens of the FAD are converted to more nucleophilic amines. If the substrate on dehydrogenation becomes suitably electron-deficient, the reduced co-enzyme will react with it. A substrate for the enzyme monoamine oxidase B is selegiline **4.18**. This propargylamine is oxidized to an ethynyliminium salt **4.19**. A conjugate addition of the dihydroflavin takes place (**4.19–4.20**).

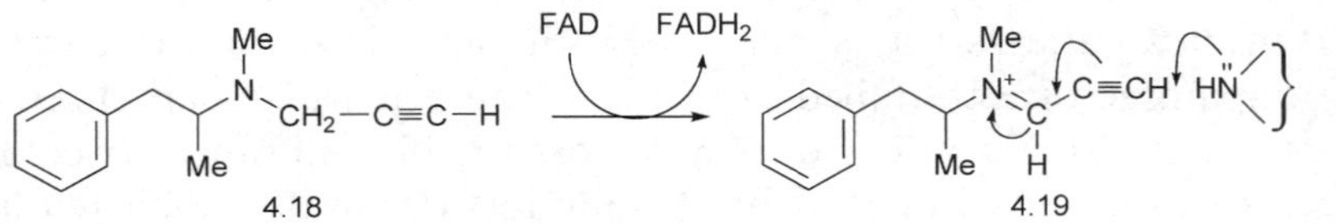

4.18   4.19

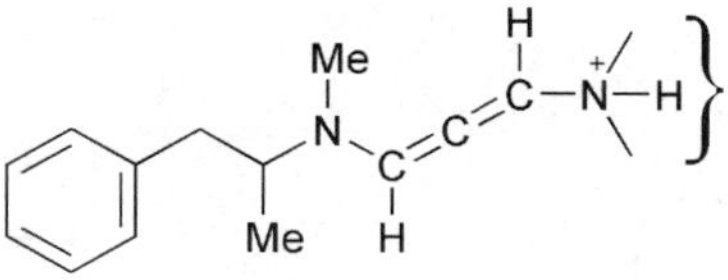

4.20

The co-enzyme has then covalently bound the substrate and it can no longer be re-oxidized to participate in a further catalytic cycle. The selegiline is a suicide inhibitor. Inhibition of the catechol *O*-methyl transferase pathway has been achieved by using catechols in which the acidity of the phenolic hydroxyl groups has been modified by attaching electron-withdrawing groups to the aromatic ring. Two examples are tolcapone, (Tasmar[R]) **4.21** and entacapone (Comtese[R]) **4.22**. These nitrocatechols are more acidic phenols and bind strongly to the catechol *O*-methyl transferase.

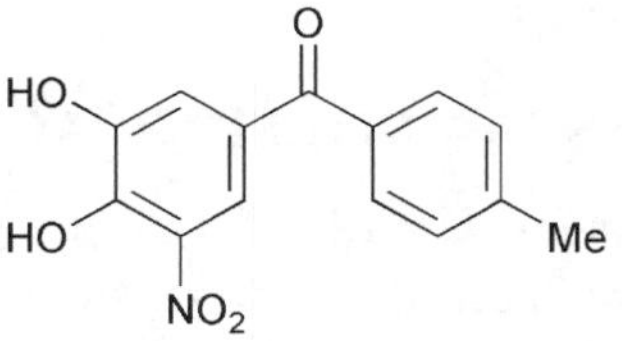

4.21 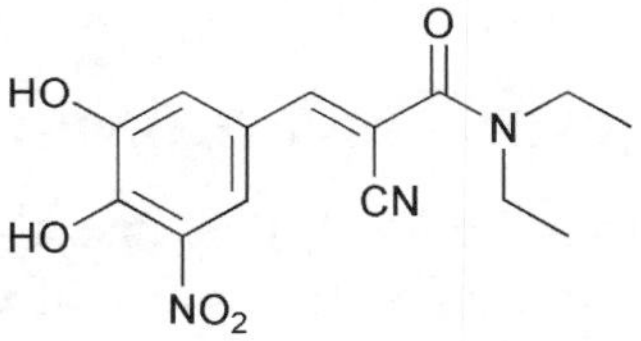   4.22

Unfortunately, the efficacy of L-DOPA therapy even when combined with these inhibitors, diminishes with time. Dopamine agonists have provided another approach to the treatment of Parkinson's disease. A series of specific dopamine receptors (D_1, D_2 and D_3) have been identified. Early dopamine agonists included an ergot derivative, pergolide **4.23** on which the more recent drug, ropirinole **4.24**, was based. The development of conformationally rigid analogues of dopamine such as dihydroexidine **4.25**, which possesses selectivity for particular dopamine receptors, is another approach. Examination of the structure of the agonist, apomorphine **4.26**, shows that it includes a conformationally rigid form of dihydroxyphenylethylamine. Furthermore, the structure is more lipophilic and as a consequence, apomorphine can reach the site of action in the brain. However, its use in the treatment of Parkinson's disease is restricted to the control of the 'on:off' episodes.

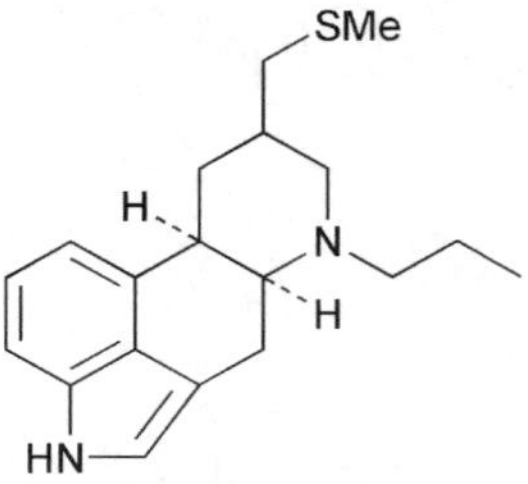

4.23 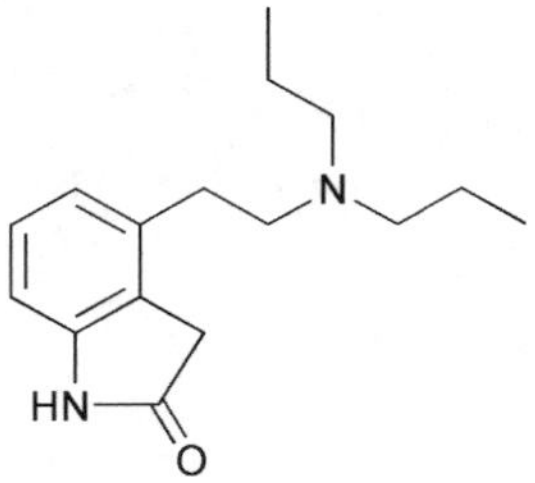   4.24

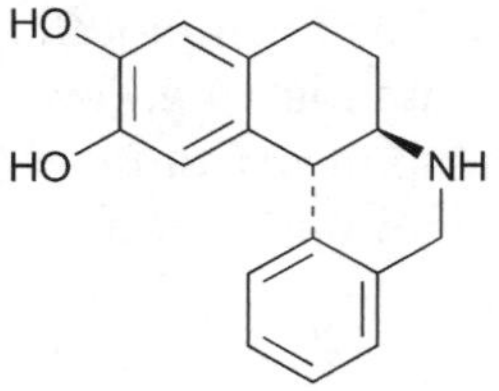

4.25 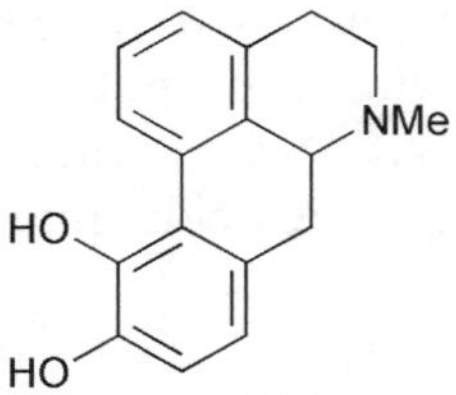   4.26

4.4   DOPAMINE ANTAGONISTS AS NEUROLEPTIC AGENTS

Whereas treatment of Parkinson's disease involved counteracting a dopamine deficiency, treatment of other mental conditions can involve controlling an overactive dopaminergic system. An overactive dopaminergic system is associated with various psychotic states. Some neuroleptic agents, which have a sedating action, are believed to act as dopamine antagonists.

   The discovery of the phenothiazine neuroleptic agents predates the discovery of dopamine as a major neurotransmitter. It was known that various anti-histamines such as diphenhydramine **4.27** had a sedative action. This property was shared by ethylenediamines such as **4.28**. The more rigid phenothiazines, **4.29** and **4.30**, were synthesized in an effort to increase the sedative action.

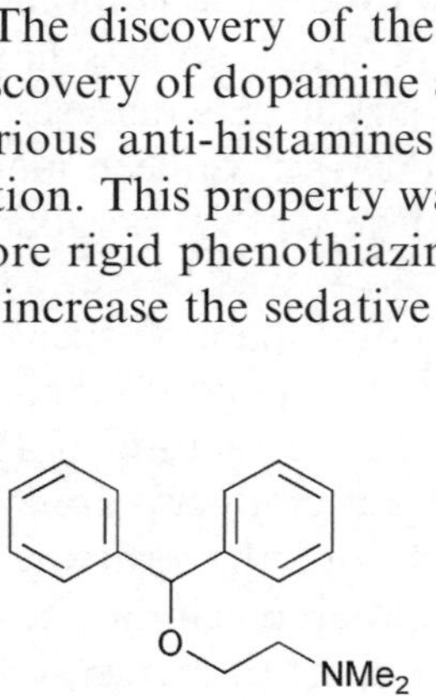

4.27 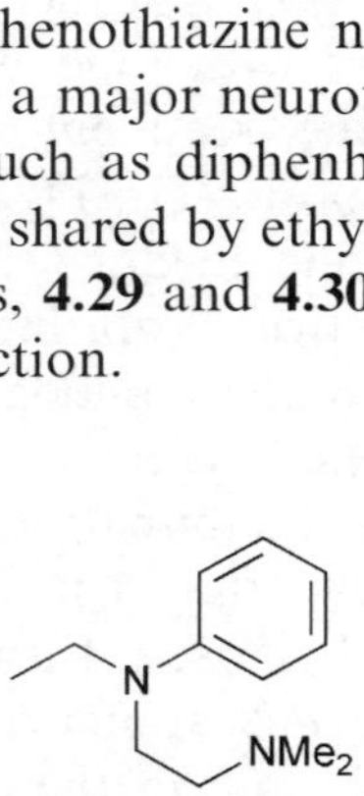   4.28 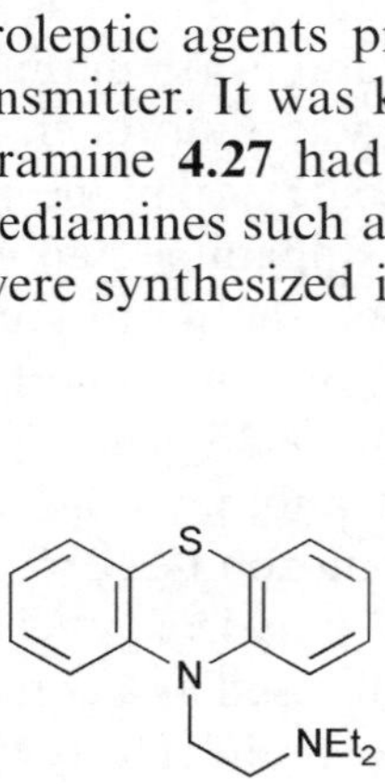   4.29

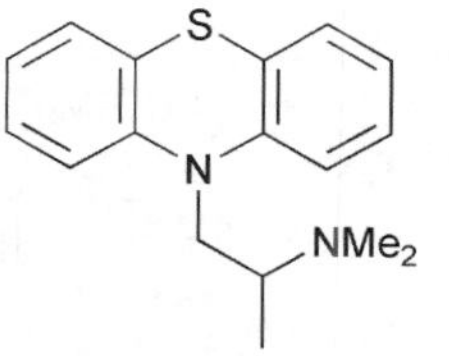

4.30 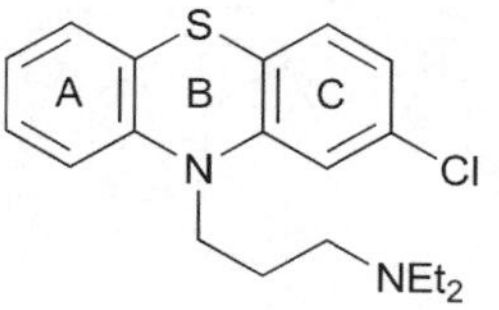   4.31 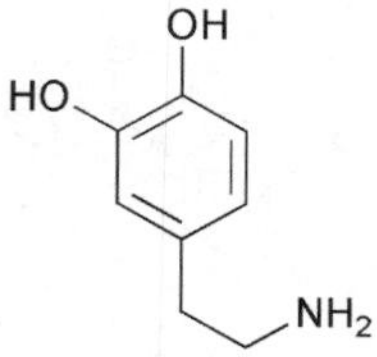   4.32

Promethazine **4.30** was found to have an anti-histamine action and to have a potentiating effect on anaesthetics. Modification of this structure in 1950 led to the discovery of chlorpromazine **4.31**, which was introduced in 1952. This compound and a number of related phenothiazines, have had a major impact on the treatment of mental illness in the relief of the agitation associated with psychoses. Chlorpromazine is now known to block the D_1 and D_2 dopamine receptors. Its structure is not planar but is folded about the nitrogen–sulfur axis. It is possible to superimpose dopamine over one ring and, given the flexibility of the side chain, the nitrogen atoms as well (see **4.32**). There is some support for this theory in that *cis*-chlorprothixene **4.33** is more active than its *trans*-isomer **4.34**. Chlorpromazine is metabolized by demethylation of the nitrogen, oxidation of the sulfur to a sulfone and by hydroxylation of one of the aromatic rings. The resultant phenol (*e.g.*, **4.35**) is conjugated with glucuronic acid and also converted to a sulfate. It can, however, cause liver damage possibly by oxidation of the phenol to a quinonimine.

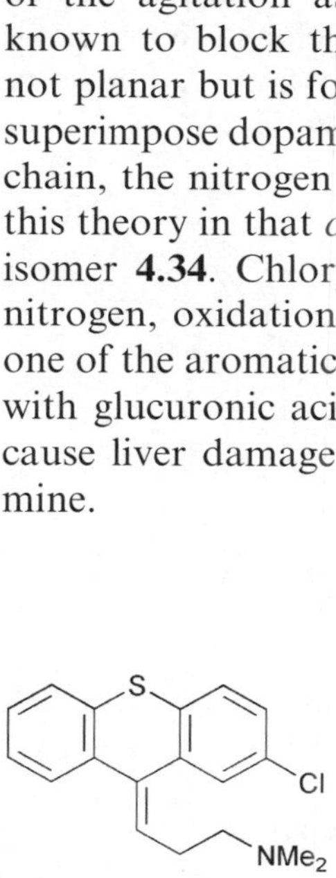

4.33 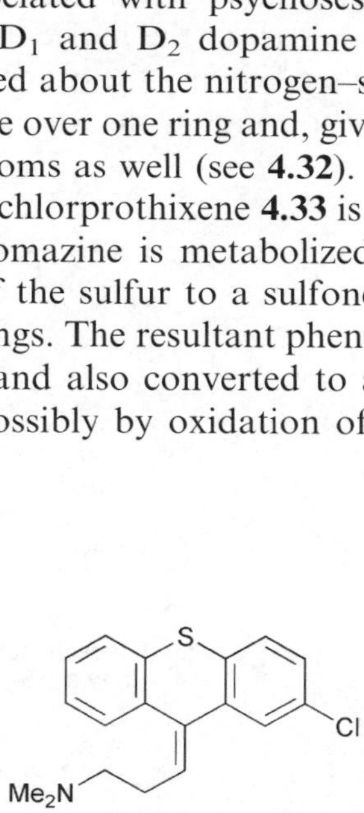   4.34 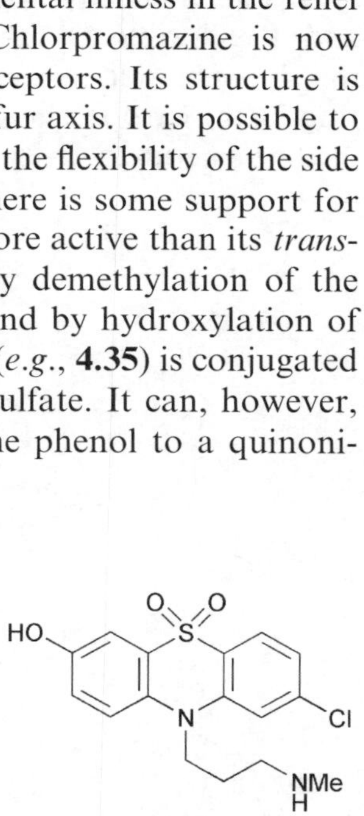   4.35

The butyrophenone, haloperidol **4.36**, has useful activity as a neuroleptic agent. Its discovery arose from attempts to increase the analgesic activity of meperidine **4.37**, a cocaine analogue. It was found that the butyrophenone analogue of meperidine **4.38** not only showed analgesic activity but also possessed a neuroleptic activity resembling that of chlorpromazine. Structural modification diminished the analgesic neuroleptic activity.

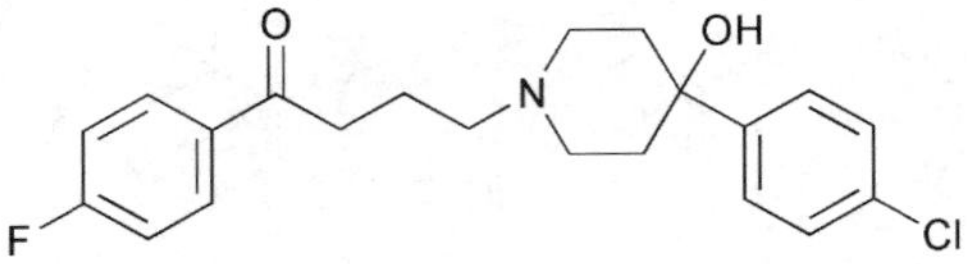

4.36 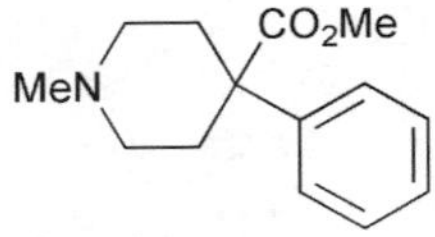   4.37

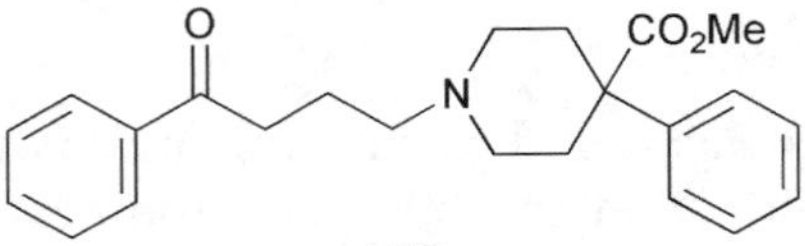

4.38

The synthesis of haloperidol **4.36** exemplifies the criterion for a good medicinal chemistry synthesis. It starts from readily available materials and it is flexible and convergent. The fact that it is convergent means that it is possible to vary one part of the molecule such as the aromatic ring or the carbon chain, while keeping the other constant. A summary of the structure:activity relationships is shown in **4.39**.

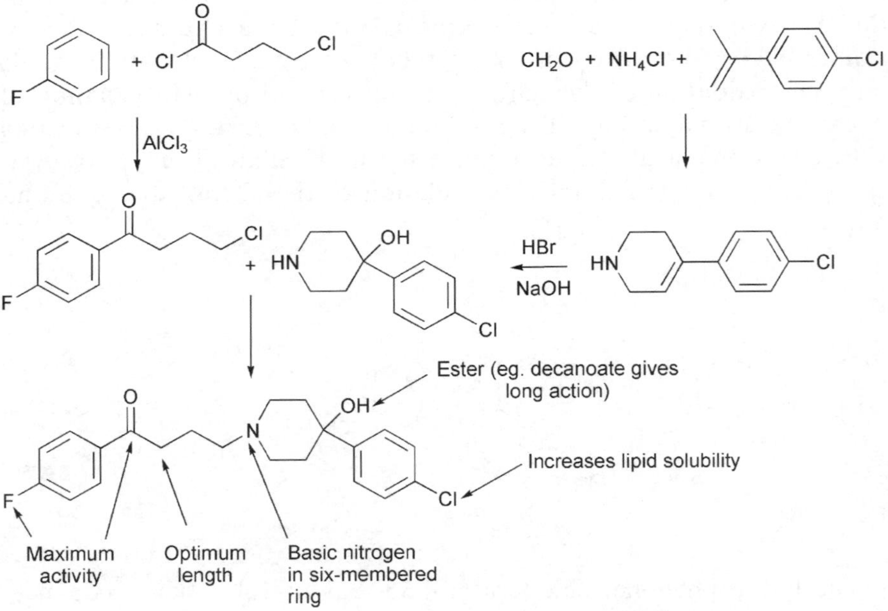

4.39

4.5   SEROTONIN AS A NEUROTRANSMITTER

Serotonin, 5-hydroxytryptamine **4.3** is one of the major neurotransmitters in the brain. Although it is found elsewhere in the body, for example in the stomach, most of the current interest on serotonin is focused on its role in the brain. It is biosynthesized from the amino acid tryptophan **4.40** by hydroxylation **4.41** and decarboxylation. The principle metabolic route involves the monoamine oxidase system, MAO-A. Serotonin is converted *via* the aldehyde, to 5-hydroxyindolylacetic acid. Another pathway involves *N*-acetylation and *O*-methylation. A product of this pathway, melatonin **4.43**, is involved in the regulation of various time-related processes in the body. The other major deactivation mechanism for serotonin is that of pre-synaptic re-uptake.

There are at least seven different 5-HT receptors and a number of subsets of these. All except 5-HT3 are G-protein-coupled receptors. The 5-HT3 receptor is linked to an ion-channel. Not surprisingly, therefore, serotonin influences a number of activities in the brain including behavioural responses such as mood, depression and sleep. Many actions in the central nervous system involve a balance between excitatory and inhibitory neurotransmitters. Whereas γ-aminobutyric acid (GABA) **4.4** is a major inhibitory neurotransmitter, serotonin **4.3** along with glutamic acid **4.5**, acetylcholine **4.6**, dopamine **4.1** and norepinephrine **4.2** are excitatory. Thus, sleep can be induced by enhancing the effect of GABA and diminishing the effect of the excitatory neurotransmitters.

4.6 THE TREATMENT OF DEPRESSION

Clinical depression is a serious mental condition, which can have its origins in the deficiency of the neurotransmitters, noradrenalin **4.2** and serotonin **4.3**. Some of the anti-depressant drugs that are used in the treatment of this disease increase the availability of these neurotransmitters.

There are several approaches to the treatment of depression. These include the use 5-HT agonists, the use of monoamine oxidase-A inhibitors and the use of serotonin re-uptake inhibitors. In the development of these agents, not only has it been important to achieve selectivity in the targets but also to develop compounds that have significant lipid solubility so that they may cross the blood: brain barrier. Furthermore, some of the compounds that have been used have multiple activities. The metabolic patterns of some drugs differ from one patient to another. For example, some patients lack the ability to metabolize paroxetine (Seroxat®) **4.44** rapidly. This has been associated with serious side effects producing an elevated level of serotonin.

Buspirone **4.45** is a 5-HT agonist, which has been used to treat depression. Buspirone has a lipophilic portion to the molecule, which enables it to cross the blood:brain barrier. Phenelzine (Nardil®) **4.46** is an anti-depressant, which functions as a monoamine oxidase-A inhibitor. Dehydrogenation produces a very reactive species that can react with the co-enzyme of the oxidase.

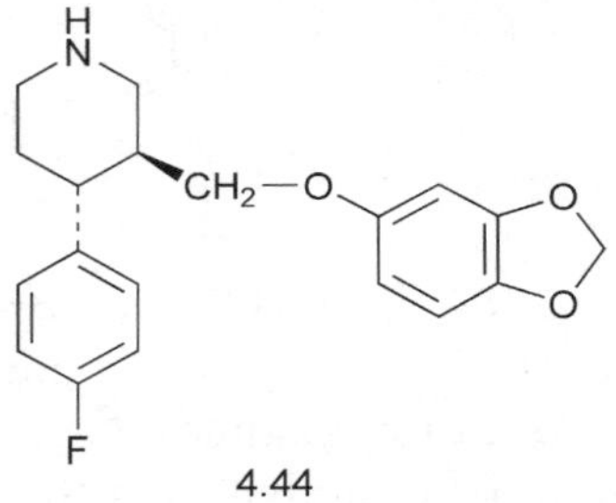

4.44

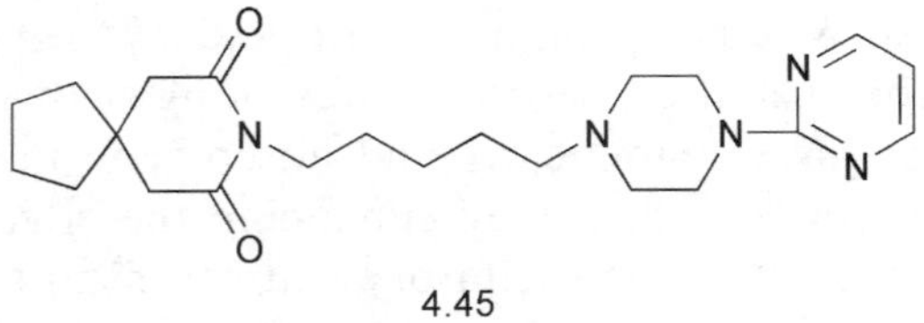

4.45

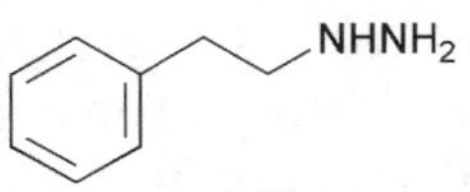

4.46

However, most attention has been directed at 5-HT re-uptake inhibitors. A series of tricyclic anti-depressants have been developed based on isosteric replacements of the sulfur of phenothiazines. These compounds act by inhibiting the neuronal re-uptake of the neurotransmitters, noradrenalin and serotonin. The compounds that were introduced first, such as imipramine **4.47**, amitryptaline **4.48** and doxepin **4.49** were relatively non-selective.

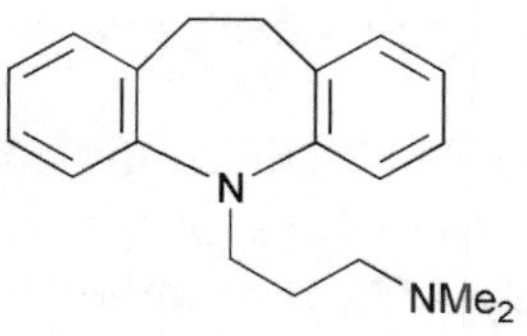

The CH_2CH_2 and CH_2O groups were iso-steric replacements for the sulfur. Because these compounds were non-selective, there were side effects arising from interactions with the adrenergic system including effects on heart rate. The metabolism of compounds such as imipramine varied between patients. Consequently, there has been a major effort to develop selective serotonin re-uptake inhibitors. SSRIs. Two compounds that are widely used are fluoxetine (Prozac®) **4.50**, introduced in 1987 and paroxetine (Seroxat®) **4.44**. Their structures are interesting since on first inspection they have more similarity to dopamine and noradrenalin. They have sufficient lipid solubility to cross the blood:-brain barrier. They are chiral molecules.

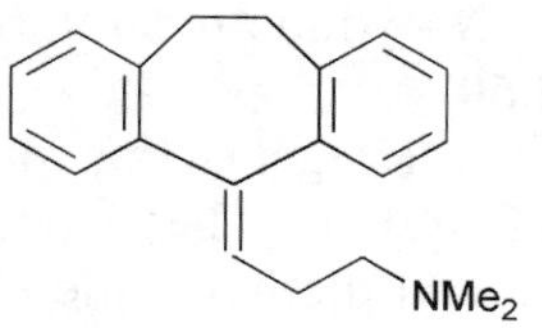

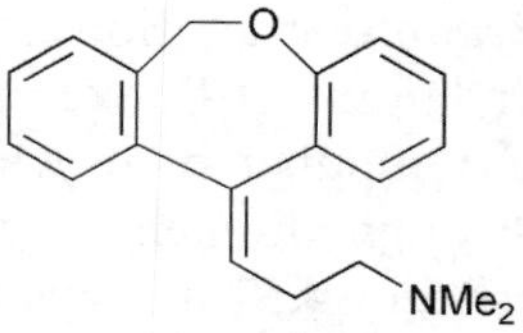

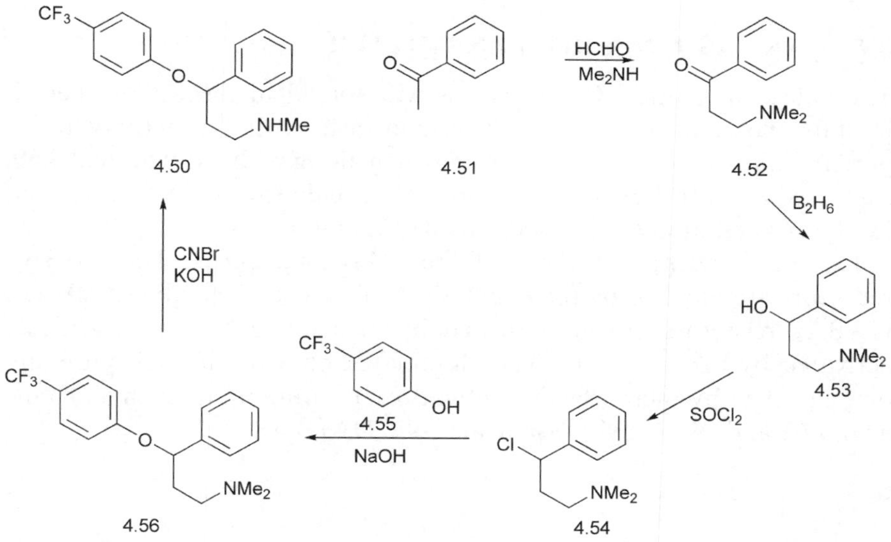

The synthesis of fluoxetine **4.50** illustrates the potential for flexibility and hence the synthesis of many analogues for structure:activity relationships. A Mannich reaction on acetophenone (**4.51–4.52**) provided one portion of the molecule. The ketone was reduced (**4.53**) and converted to chloride (**4.54**). A nucleophilic displacement of the chloride with 4-trifluoromethylphenol led to the aryl ether (**4.56**). Finally, a von

Braun degradation of the *N,N*-dimethylamino group of **4.56** gave methylamine of fluoxetine **4.50**.

A number of psychomimetic agents that produce hallucinations act
through the central 5-HT system. These include the drugs of abuse,
psilocin **4.57** and lysergic acid diethylamide (LSD) **4.58**. A number of
other indole derivatives are used to treat CNS problems, *e.g.*, sumatrip-
tan for migraine, and possibly act *via* the 5-HT system.

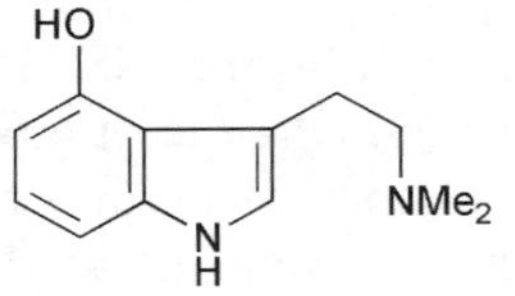

4.57 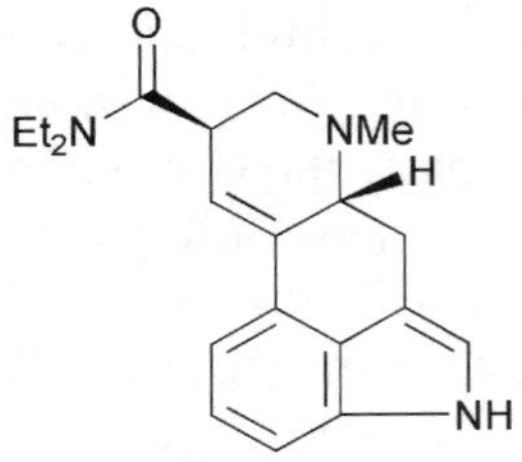   4.58

4.7   GABA AS A NEUROTRANSMITTER

The neurotransmitter, GABA **4.4** is widespread in the CNS, where it
inhibits the central neurones. This contrasts with the activity of its
parent amino acid, glutamic acid **4.5** which along with aspartic acid **4.59**,
excites the central neurones. There is a balance of inhibitory and
excitatory action involving these related amino acids.

There are several sub-types of the GABA receptors. The GABA$_A$
receptors are blocked by the convulsant substance bicuculline **4.60**. The
GABA$_B$ receptors are not affected by bicuculline but are selectively
activated by baclofen **4.61**. The glutamate and aspartate receptors are
distinguished by their selective activation by compounds such as kainic
acid **4.62** and *N*-methyl-D-aspartate (NMDA) **4.63**.

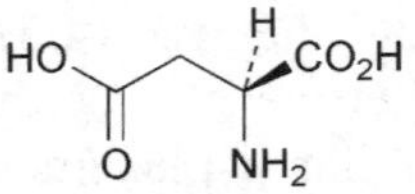

4.59 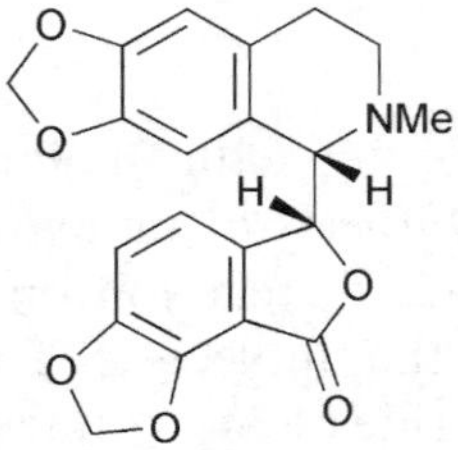   4.60

4.61   4.62   4.63

The GABA$_A$ receptors belong to the family of ligand-gated ion-channel receptors. When a drug binds to the GABA$_A$ receptors and activates it, the chloride ion-conductance through an ion-channel is increased and the concentration of chloride ion in the cell increases.

4.8   THE TREATMENT OF EPILEPSY

Epilepsy is a widespread disease which includes a number of different seizure disorders that can recur. The common feature is the sudden, excessive and disorderly discharge of cerebral neurones. Convulsions may occur if the motor cortex is involved. Seizures have been divided into two groups, partial and general, depending on the extent of the brain that is involved. Many epileptics have a severe impairment of the GABA system. Drug therapy is the most widely used method of control and a number of the common drugs interact with different facets of the GABA system. The compounds in use may modify the biosynthesis of GABA, diminish its metabolism, act as agonists and modify the chloride ion-channels. The action of the drugs is to enhance the effect of GABA and to diminish the response to the excitatory glutamate.

Progabide **4.64** is a GABA pro-drug and serves, by hydrolysis of the imine and the amide, to augment the amount of GABA that is available. The benzophenone moiety provides the lipid solubility to enable the drug to cross the blood:brain barrier. The phenolic hydroxyl group provides neighbouring group participation to facilitate the hydrolysis. The fluorine atom prevents hydroxylation para to the phenol and thus oxidation to a quinone. The amide is a neutral derivative of a carboxylic acid. On the other, vigabatrin **4.65** is an irreversible inhibitor of GABA transaminase and hence it inhibits GABA metabolism. Sodium valproate **4.66** leads to a reduced degradation of GABA by blocking succinic dehydrogenase and hence by feedback regulation, it increases the concentration of GABA.

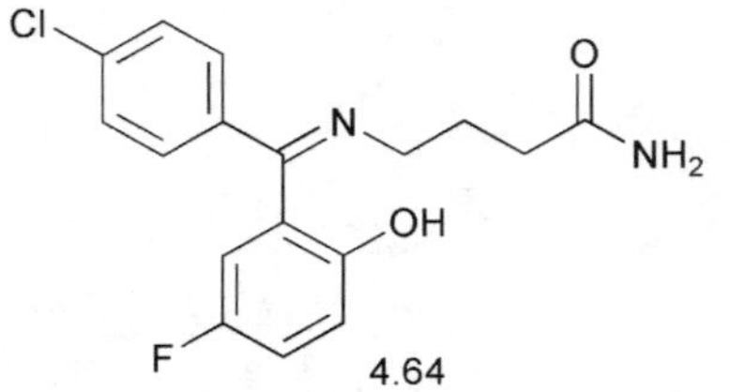

4.64 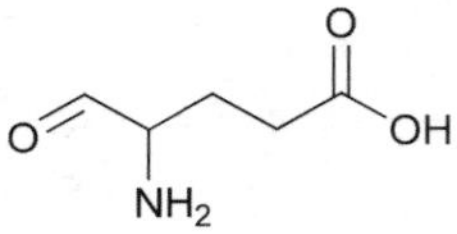   4.65 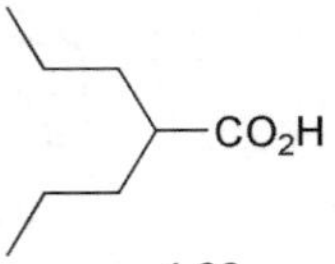   4.66

Lamotragine **4.67** acts by inhibiting the release of the excitatory glutamic acid, and hence it acts to facilitate the action of the limited amount of GABA that is available to an epileptic.

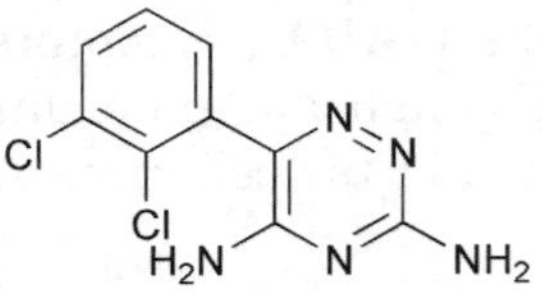

4.67 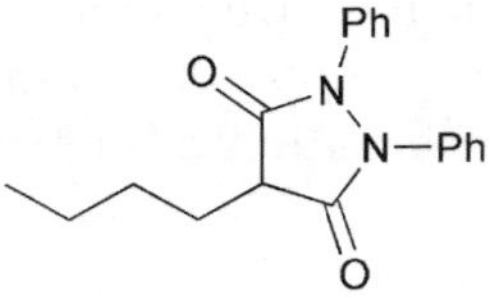   4.68 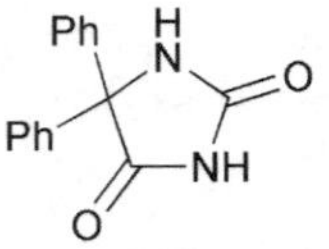   4.69

Phenylbutazone **4.68** and phenytoin **4.69** are older drugs, which have a direct action on the chloride ion-channel and lead to its opening. The BZDs have an allosteric enhancement of GABA efficacy and are discussed later.

Phenobarbital (phenobarbitone) **4.70** was synthesized as a hypnotic in competition to barbital (5,5-diethylbarbituric acid) and fortuitously it was discovered that it possessed anti-convulsant activity. A series of hydantoins were then synthesized as analogues including diphenylhydantoin (phenytoin). Other successful analogues that followed were trimethadione **4.71** and cyclic amides such as primidone **4.72**. The latter is a pro-drug for phenobarbital.

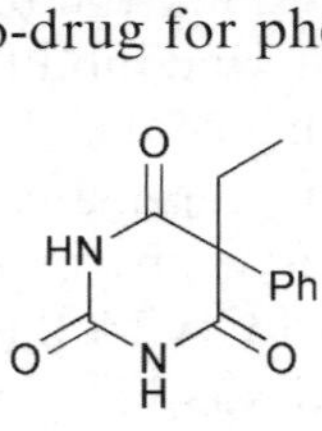

4.70 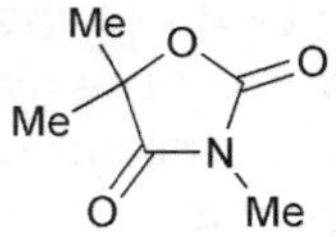   4.71 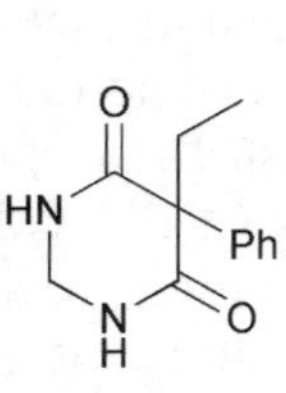   4.72

4.9   BENZODIAZEPINES AS ANXIOLYTIC AGENTS

The treatment of anxiety and sleep disorders is dominated by the BZDs. These drugs induce sleep (act as hypnotics) in high doses and lead to sedation and reduced anxiety (anxiolytic) at lower doses. They function by an enhancement of the GABA-mediated inhibition of the CNS and act in an allosteric manner on the GABA receptor. The discovery of the

BZDs was both fortuitous in the nature and the timing of their discovery.

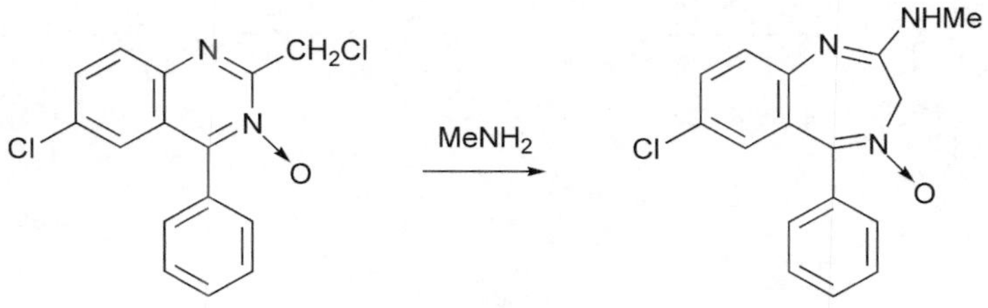

4.73   4.74

A number of animal screens for tranquilizers involving relaxing muscle in the cat and foot shock tests on mice were available in the 1950s. Today such screening tests would be replaced by receptor and enzyme screens. In 1955, a series of quinazoline derivatives were being examined. Treatment of the quinazoline *N*-oxide **4.73** with methylamine gave an unknown compound, which was not tested until the project was coming to an end in May 1957. The compound showed activity comparable to the well-known tranquilizers phenobarbital and chlorpromazine. Chemical studies then established the structure as a benzodiazepine **4.74**. The patent was filed in May 1958 and granted in July 1959 and the compound was marketed in 1960 as chlordiazepoxide (Librium®). This rapid timescale is very different from that which occurs today and reflects the caution with which new medicines are now introduced.

Studies on related compounds showed that not all the structure was required and activity was found with diazepam (Valium®) **4.75**. This led to many other BZDs including temazepam **4.76** and nitrazepam **4.77**.

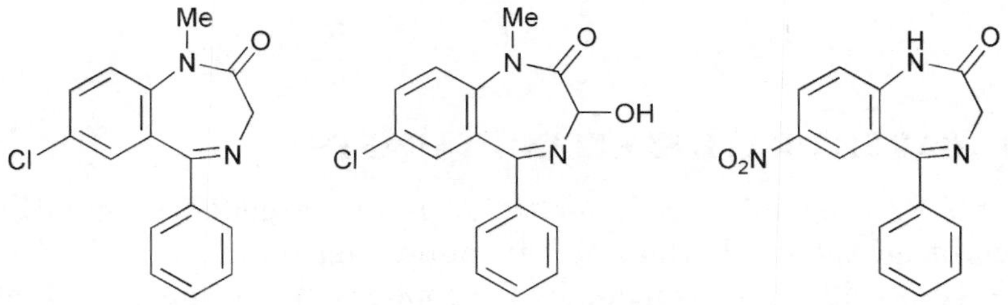

4.75   4.76   4.77

The synthesis of the BZDs, *e.g.*, **4.75**, from an aminobenzophenone **4.78** *via* the chloro- and amino-acetyl derivatives **4.79** and **4.80** and the cyclic amide **4.81**, fulfils the criterion for a useful medicinal chemistry synthesis in terms of its flexibility. Many structural variations have been synthesized with different substituents on the aromatic rings and with different units attached to the seven-membered diazepine ring.

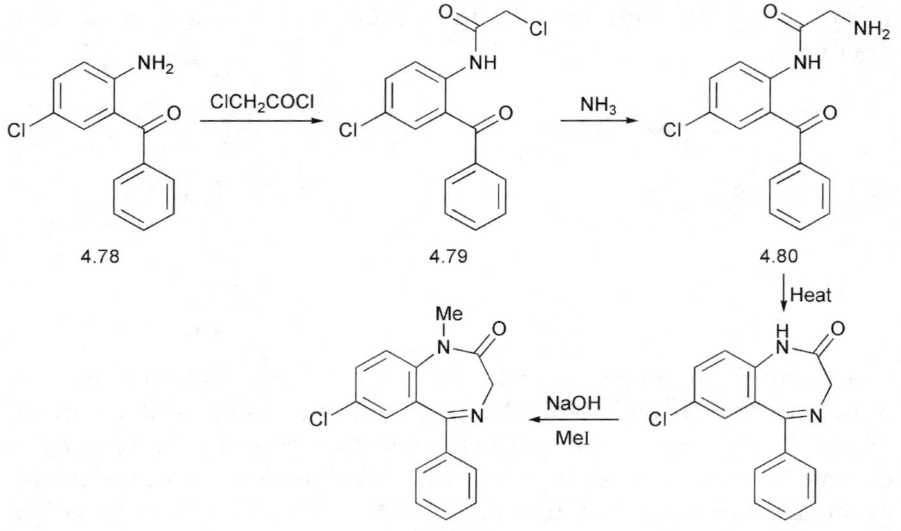

The BZDs function by binding in an allosteric manner to the $GABA_A$ receptor close to the GABA binding site. This binding alters the shape of the receptor so that the effect of GABA is changed. The consequence of this is an enhanced and in some instances longer opening of the ion-channel.

4.10   BARBITURATE SLEEPING TABLETS

The $GABA_A$ ligand-gated ion-channel is the target for a number of drugs including the barbiturate hypnotics. Barbituric acid itself is a vinylogous acid, which is made by the condensation of urea and diethyl malonate. Substituted malonate esters such as diethyl ethyl-phenylmalonate **4.84**, prepared from ethyl phenylacetate **4.82** *via* **4.83** afford the barbiturates, *e.g.*, **4.70**. Among the best known are pheno-barbital (phenobarbitone) **4.70**, amobarbital **4.85** and thiopental **4.86**. The increased lipophilicity associated with the introduction of the sulfur and the alkyl chains allows thiopental to cross the blood:brain barrier more easily and it thus affords a rapid intravenous anaesthetic.

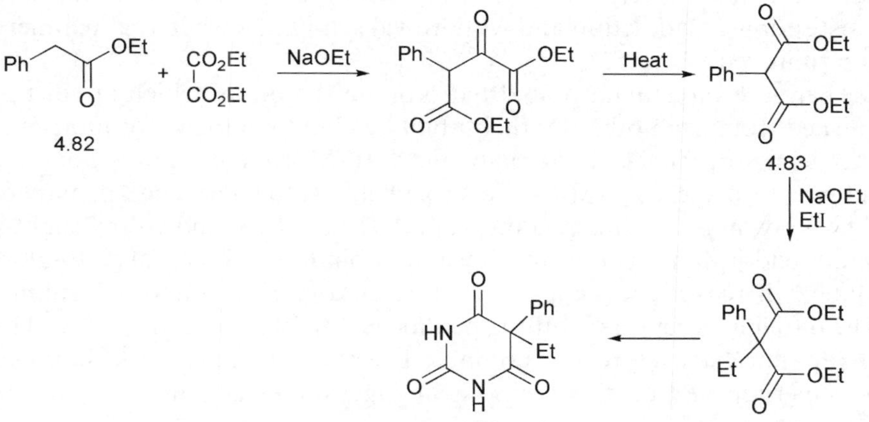

4.82 4.83 4.84 4.70

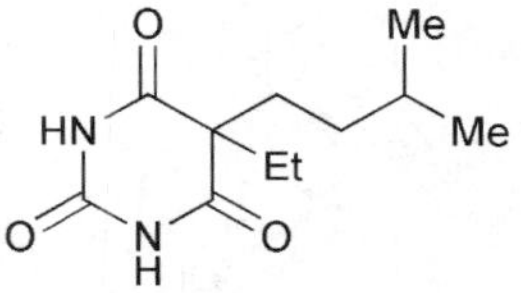

4.85

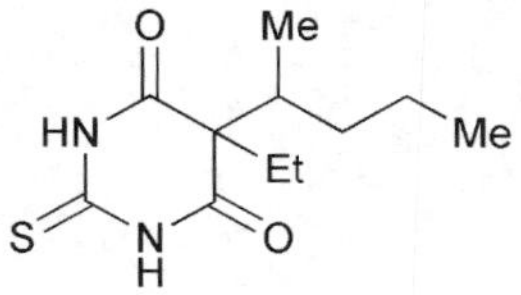

4.86

4.11 OPIOIDS AS ANALGESICS

Opium, from the poppy *Papaver somniferum*, is one of the oldest drugs known to man. Its pain relieving and euphoric properties were known to the Ancient Greeks. Morphine **4.87** was first isolated from opium in 1803 and its structure was established in 1925. Its synthesis was reported in 1952 and the X-ray crystal structure was determined in 1968. It co-occurs with codeine **4.88** and thebaine **4.89** together with a family of benzylisoquinoline alkaloids.

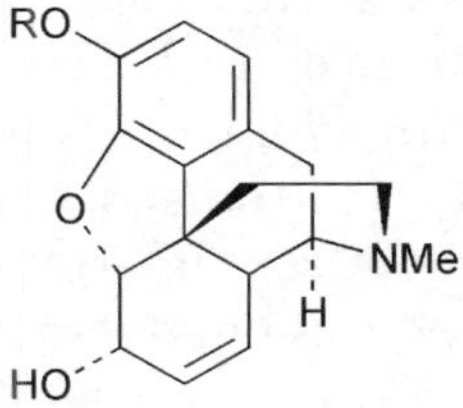

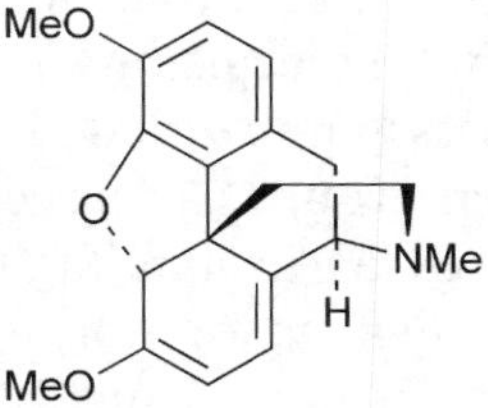

4.87 R = H
4.88 R = Me

4.89

Despite its serious side effects, morphine is still used for the treatment of chronic pain particularly in terminal diseases. The side effects include euphoria, the depression of the respiratory centres, constipation and most seriously, addiction and withdrawal symptoms when the treatment is terminated.

There are three main opioid receptors in the brain, which are distinguished from each other by their ability to bind analogues of morphine such as nalorphine **4.90** and naloxone **4.91**. Morphine is not the natural substrate for these receptors. This role belongs to some small pentapeptides known as the enkephalins, *e.g.*, **4.92** and **4.93**, and to the slightly larger endorphins. The μ (mu) receptors bind morphine but not nalorphine, whereas the κ (kappa) receptors bind morphine and nalorphine. The δ (delta) receptors bind the peptide enkaphilins and morphine. The enkaphilins have a tyrosine terminus. The phenolic ring of morphine can be superimposed on this amino acid suggesting that both bind in the same way to the receptors.

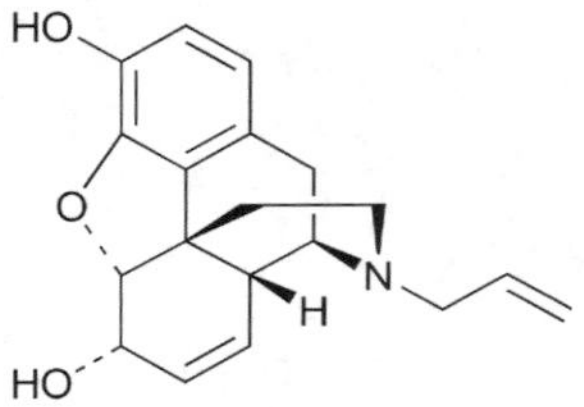

4.90 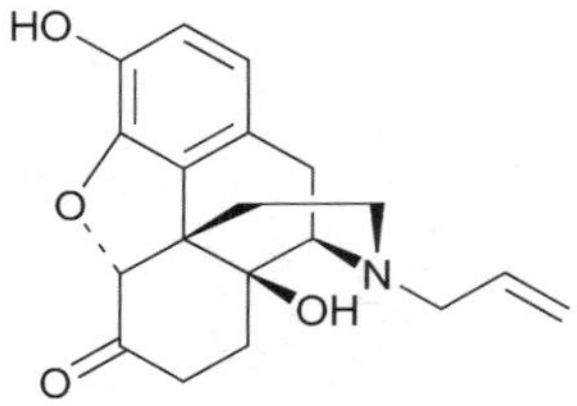   4.91
H–Tyr–Gly–Gly–Phe–Leu–OH   H–Tyr–Gly–Gly–Phe–Met–OH
4.92   4.93

The availability of morphine and many synthetic analogues have led to major investigations into structure:activity relationships. Many of these studies have been directed at trying to separate the analgesic pain-relieving properties from the addictive properties. These studies can be grouped into four main areas. First, the examination of the role of the functional groups of morphine; second, the simplification of the carbon skeleton; third, the creation of more rigid bulky structures and finally the synthesis of antagonists. Although at first sight morphine may appear to be a flat molecule, this is far from correct. The X-ray structure shows it to possess an inverted 'T' shape **4.94**.

The functional groups can affect both the transport and metabolism of a drug as well as its binding to the receptor. Methylation of the phenolic hydroxyl group of morphine reduces but does not eliminate the analgesic effect. In fact, demethylation occurs in the liver and some codeine is converted to morphine. The phenolic hydroxyl group is important for activity. When the alcoholic hydroxyl group is methylated as in heterocodeine, the product is quite active as an analgesic. Diacetylmorphine (heroin) is very active. 6-Ketomorphine shows activity.

In each case, the changes to these groups facilitate the transport of morphine across the lipid blood:brain barrier. Hydrolysis or reduction can then take place in the brain. Reduction of the double bond to give dihydromorphine retains the activity. *N*-Demethylation leads to reduced activity, while replacement of the methyl group by ethyl, propyl and butyl leads to a reduction in activity. Suprisingly, there is then a rapid increase in activity and a phenylethyl substituent is very active suggesting that there may be an additional hydrophobic binding site in the receptor into which the larger alkyl group fits. Removal of the nitrogen containing bridge led to a complete loss of activity.

Morphine was synthesized as a racemic mixture and the enantiomers were resolved. Not suprisingly only one enantiomer had analgesic activity.

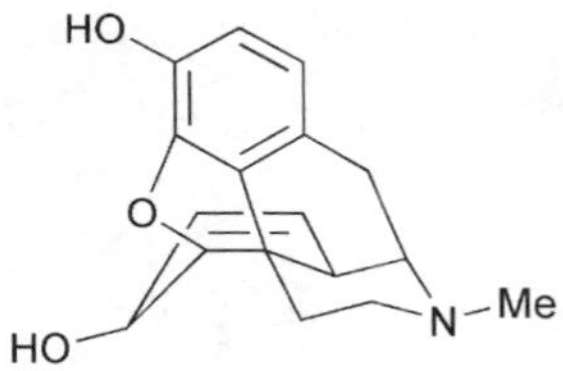

| 4.94 | 4.95 | 4.96 |

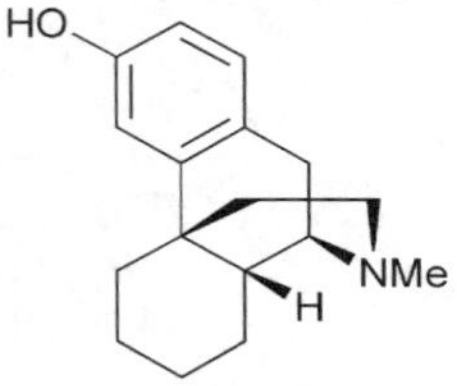

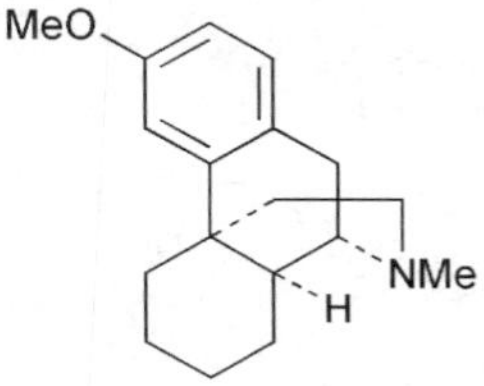

Variations in the skeleton involving the stepwise removal of the rings gave some useful information. In practice, most of these compounds were made by total synthesis. The morphinan series lack the ether bridge. Levorphanol **4.95** has useful activity. The methyl ether of the enantiomer, dextromethorphan **4.96** has no analgesic or addictive activity but it suppresses coughs and it is used in many 'over-the-counter' cough mixtures. This is a good example of enantiomers having different biological activities.

Removal of ring C gave the benzomorphan series. Metazocine **4.97**, R = Me, was quite active and phenazocine **4.97**, R = CH_2CH_2,Ph was very active but there was some drug dependence.

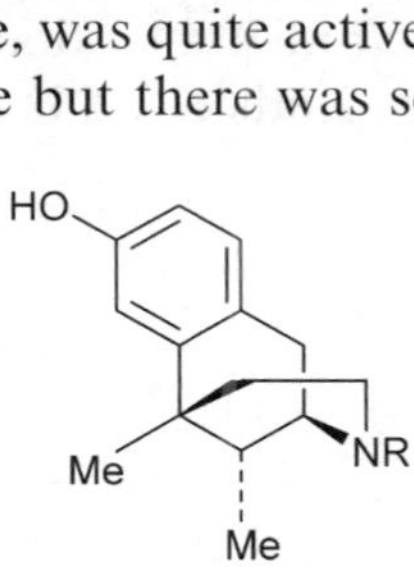

| 4.97 | 4.98 | 4.99 |

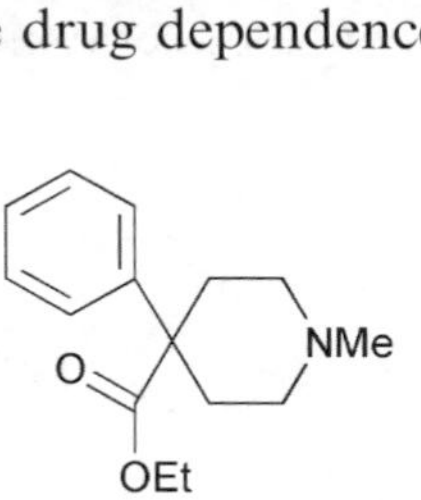

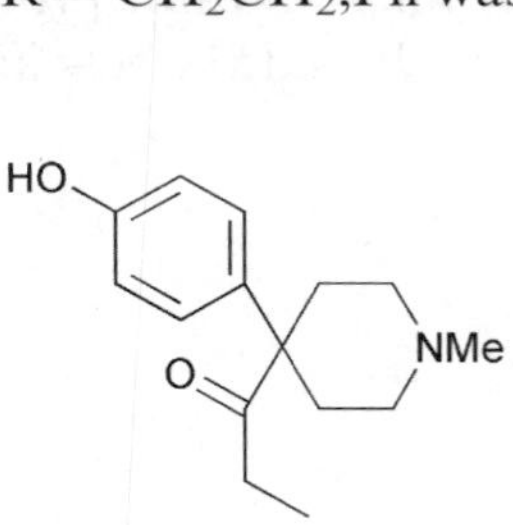

Studies on cocaine analogues gave a series of compounds that turned out to have analgesic activity related to that of the opioids. These

4-phenylpiperidines, although originally conceived as cocaine analogues, can now be viewed as analogues of morphine. They include
pethidine (meperidine) **4.98** and ketobemidone **4.99**. A very lipophilic
analogue of pethidine, which was introduced in 1960, is fentanyl **4.100**.
It is used for pain relief and surgical anaesthesia. Fentanyl is prepared by
reacting *N*-phenylethylpiperidone with aniline and then reducing the
imine to give 4-anilino-*N*(phenylethyl)piperidine. Reaction of this with
propionyl chloride gave fentanyl.

A compound that binds to the morphine receptors and has morphine-
like effects and is orally active with rather less serious side effects is
methadone **4.101**.

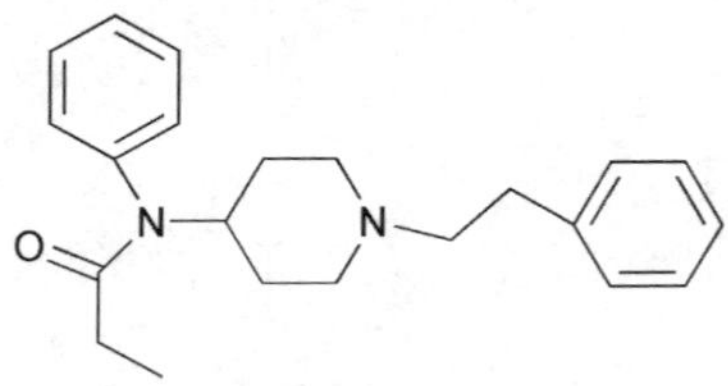

4.100 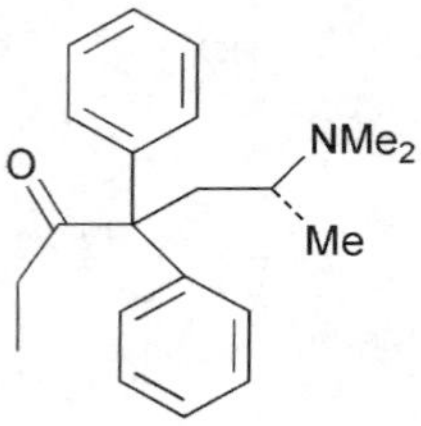   4.101 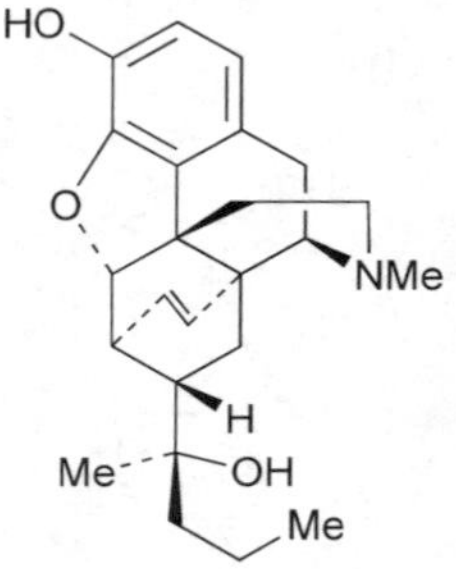   4.102

Another approach to selectivity is to make the structure more rigid
and more bulky. This can be achieved by increasing the number of rings.
The oripavines are derived by a Diels–Alder reaction from the diene,
thebaine **4.89** and methylvinyl ketone. Etorphine **4.102** in this series is
quite lipophilic and has powerful analgesic properties with limited side
effects. Hence, quite significant separations of analgesia and addiction
can be made.

The development of opioid antagonists provides an approach to the
addiction problem. Nalorphine **4.91** is an opioid antagonist. Not only
does it bind and not produce the analgesic effect, but it also does not
produce the side effects and hence the craving that leads to addiction.

CHAPTER 5

Local and Circulatory Hormone Targets

5.1 AIMS

The aim of this chapter is to describe the medicinal chemistry of compounds that modify the action of local and circulatory hormones by interacting with their biosynthesis, metabolism and binding to receptors. By the end of this chapter you should be aware of:

- the role of histamine as a local hormone and the use of histamine antagonists in the treatment of peptic ulcers;
- the role of aspirin and other non-steroidal anti-inflammatory agents as inhibitors of prostaglandin biosynthesis; and
- the various roles of the steroid hormones and their structural modification in the development of anti-inflammatory steroids and the oral contraceptives.

5.2 INTRODUCTION

In the previous two chapters we have considered the medicinal chemistry of neurotransmitters and their functions. Another class of hormones is the circulatory hormones, which are formed in one gland and translocated in the circulatory system to a target organ. These include the peptide hormones and the steroids. A further group are the local hormones, which are produced and exert their effect locally, often within the same tissues. These are exemplified by histamine and the prostaglandins. A chemical signalling substance that also falls into this area is the gas nitric oxide. In this chapter we are concerned with the medicinal chemistry of local and circulatory hormone targets.

5.3   HISTAMINE AS A TARGET

Histamine **5.1** is a widespread local hormone. It affects the circulatory system producing arteriolar dilation and increases capillary permeability. The local inflammation and swelling around an insect bite are due to the release of histamine. In the lungs it produces bronchial constriction while in the stomach it stimulates gastric acid secretion. There are four families of receptors of which the H_1, H_2 and H_3 are the best known. There is some evidence for the existence of an H_4 receptor.

5.1   5.2

   Among the classical H_1 antagonists are compounds such as mepyramine **5.2** and diphenhydramine **5.3**, which are used in over-the-counter medicines to reduce inflammation and in cough mixtures to reduce bronchial constriction (Benadryl). A typical 'over-the-counter' cough mixture contains diphenhydramine hydrochloride as an anti-histamine and dextromorphan or pholcodine (the morpholinoethyl ether of morphine) as a cough suppressant. Other preparations can contain paracetamol as an anti-pyretic, diphenhydramine as an anti-histamine and pseudoephedrine as a nasal decongestant. Since diphenhydramine is quite lipophilic, it can cross the blood:brain barrier. It has a marked sedative action, a side effect that has been exploited in some mild sleeping tablets. Diphenhydramine is prepared by the nucleophilic substitution of the bromine of diphenylmethyl bromide with 2-dimethylaminoethanol. The aromatic component is easily prepared from benzophenone and the 2-dimethylaminoethanol is formed from dimethylamine and ethyleneoxide. This simple synthetic route has sufficient flexibility to allow the preparation of a range of compounds.

5.3   5.4

5.4 HISTAMINE ANTAGONISTS IN THE TREATMENT OF PEPTIC ULCERS

H_2 Antagonists have been thoroughly studied in the context of the control of peptic ulcer disease. The development of anti-ulcer drugs based on histamine is one of the classical case histories of medicinal chemistry. The parietal cells in the stomach release acid under the stimulation of the peptide hormone, gastrin, and the local hormones, histamine **5.1** and acetylcholine. Stomach ulcers are formed partly as a result of trauma but also because of infection by a bacterium, *Helicobacter pylori*. If an ulcer has formed in the stomach lining, the presence of a strongly acidic environment inhibits the healing process. The development of selective histamine antagonists therefore provided a means of reducing the acidity of the stomach and allow the ulcer to heal.

The first generation of drugs was developed by the Smith Kline French Laboratories starting in 1964. The first step in designing an inhibitor was to establish a reliable bioassay. In this case it was to measure the pH of a rat's stomach after treatment with the test compound. The chemical strategy was to take the structure of histamine and to retain the imidazole ring as one binding group and to modify the side chain amine. The first lead compound came from a series in which the basicity and hydrogen-bonding ability of the side chain amine was modified by converting it to a guanidine and by increasing the length of the side chain as in **5.4**. Further modification of the chain length and conversion of the side chain to a thiourea gave improved activity. The guanidine side chain was strongly basic and hence like the side chain of histamine, it would be protonated at physiological pH. Structural changes were then made to retain a polar group in the side chain, which was capable of hydrogen-bonding interactions but which was relatively non-basic, *i.e.* the compound might bind but not have any biological activity. A selective competitive H_2 antagonist, burimamide **5.5** was obtained. It then became a case of fine-tuning the structure to obtain a suitable drug.

The imidazole ring is a tautomeric structure **5.6**. However, when considering binding to a receptor, one tautomer could well be favoured for binding. The hydrogen bonding capabilities of the two tautomers are quite different. It was found that introducing a methyl group onto the imidazole ring and an electron-withdrawing group in the side chain favoured the required tautomer and conformation. Isosteric replacement of a methylene by a sulfur in the side chain gave an enhancement of activity. The product, metiamide **5.7**, reached clinical trials.

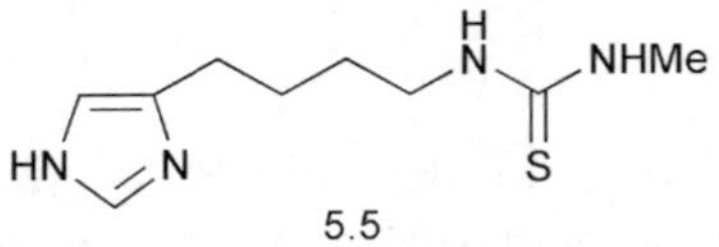

A side effect of this compound was noticed in a small number of patients. There was a decrease in the white blood cell count, a condition known as granulocytopenia. This effect was associated with the thione in the side chain. This functional group was replaced by a cyanoimine, giving cimetidine (Tagamet®) **5.8**. This proved to be a major development in the control of peptic ulcers.

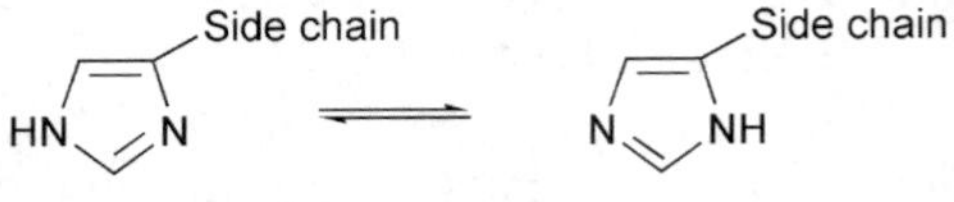

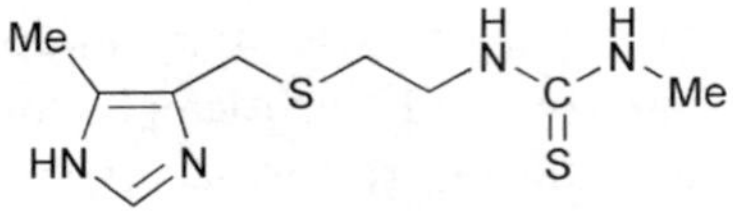

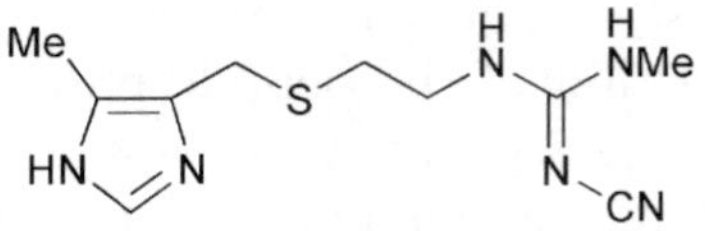

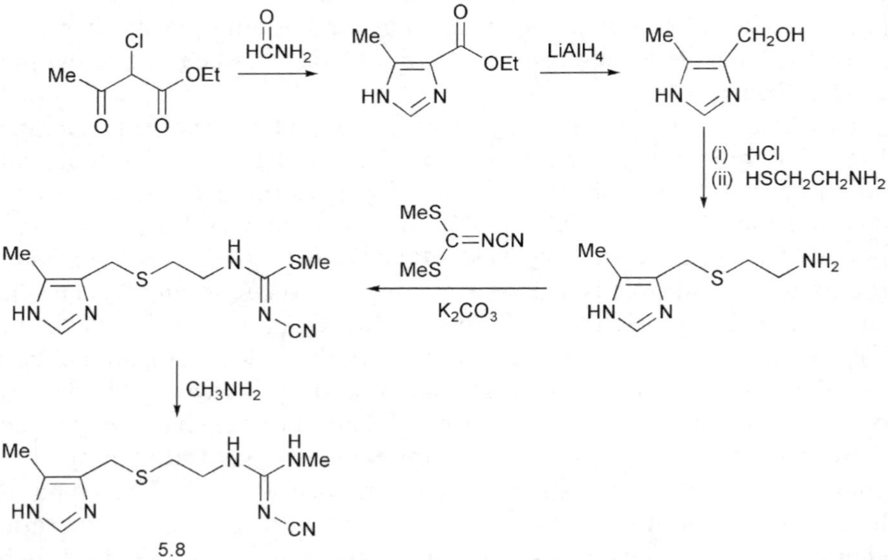

Cimetidine **5.8** is synthesized from ethyl 2-chloroacetoacetate by a route that has been adapted for large-scale production.

An interesting feature of this work is the time-scale. The research programme began in 1964 and the first leads were obtained in 1968. Burimamide was developed in 1970, metiamide in 1972 and cimetidine in 1973. After trials and drug registration and approval processes were complete, cimetidine was released in 1976 in the UK and in the USA in

1977. A period of 12–13 years had elapsed between the start of the programme and the first financial return.

The second-generation drug, ranitidine (Zantac®) **5.12**, appeared in 1981 and by 1986 it had become a major drug. The group at the Allen and Hanbury (Glaxo) laboratories that developed this drug took burimamide **5.5** as their starting point, which by 1972 was in the literature. Among other changes, modification of the imidazole ring gave a 5-aminotetrazole derivative **5.9**, which was equipotent to burimamide. The aminotetrazole was less basic than the imidazole of histidine and yet retained activity. A decision was then taken to replace the tetrazole ring by a more readily accessible heterocycle, in this case a furan. The butyl side chain was replaced by a methylthioethyl chain. 2-Furfurylmercaptan was commercially available and the furan analogue **5.10** was prepared. However, this compound was poorly soluble and gave variable results. The solubility problem was overcome by introducing a dimethylamino group **5.11** that would form a salt. This could be introduced by a Mannich reaction. The thione of metiamide was by this time known to be the cause of side effects in the first generation of drugs. Replacement by a cyanoimine as in cimetidine, gave poorly crystalline material, which was difficult to handle and hence the nitrovinyl group was introduced to give ranitidine **5.12**. This compound was first prepared in the summer of 1976 and launched as Zantac® in 1981.

5.9

5.10

5.11

5.12

A different approach to the control of gastric acid secretions has been to develop inhibitors of the H^+/K^+ ATPase enzyme system on the surface of the gastric parietal cells. This system is responsible for the secretion of acid by producing protons in exchange for potassium ions and is known as a proton pump. Omeprazole **5.13** was approved in 1988 for use as a proton pump inhibitor.

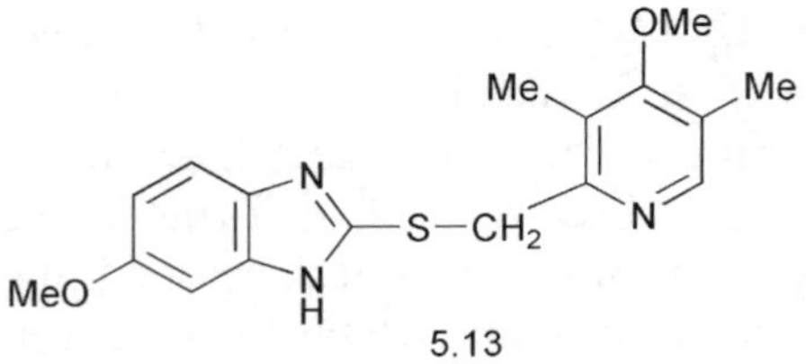

5.13

5.5   THE PROSTAGLANDINS AND NON-STEROIDAL ANTI-INFLAMMATORY AGENTS

The prostaglandins are a widespread group of local hormones, which mediate a range of biological processes including inflammation. Their biosynthesis is the target for a number of drugs including aspirin **5.14**.

Aspirin **5.14** is the best known of a family of drugs that are known as the non-steroidal anti-inflammatory drugs (NSAIDs). These drugs are used because of their analgesic (pain-killing), anti-inflammatory and anti-pyretic (fever reducing) properties. Aspirin has its origins in folk medicine. Chewing the bark of the willow tree *(Salix europea)* alleviated pain associated with rheumatism, toothache and headache. Salicin **5.15** and salicylic acid **5.16** were originally isolated from this source. They were used in the nineteenth century for the treatment of rheumatic fever and for their anti-pyretic and anti-inflammatory properties. Salicylic acid became readily available from phenol **5.17** by the Kolbe reaction. However, it produced side effects involving gastrointestinal damage. It was found that acetylation reduced these side effects and aspirin was introduced in 1899. Other salicylates derived from natural sources have useful anti-inflammatory action. Oil of wintergreen (methyl salicylate) is used in the topical treatment of sprains and local inflammation. The presence of the methyl ester gives it greater lipid solubility and thus methyl salicylate can cross fat barriers on the skin and is more easily absorbed. Some salicylates that have been introduced produce fewer side effects. These include the dimer salsate **5.18** and diflunisal **5.19**.

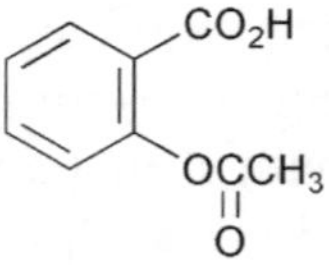

5.14 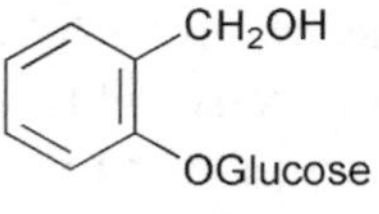   5.15 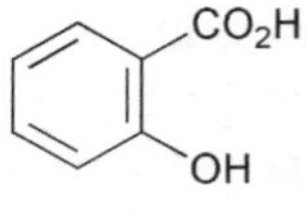   5.16

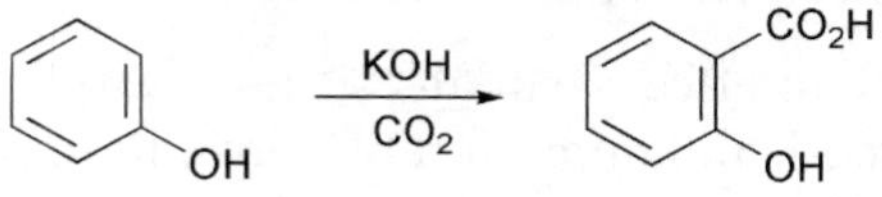

5.17 → 5.16 (KOH, CO_2)

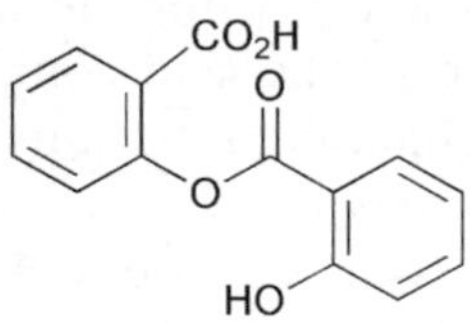

5.18 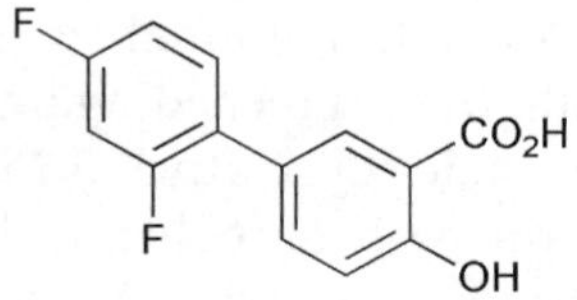   5.19 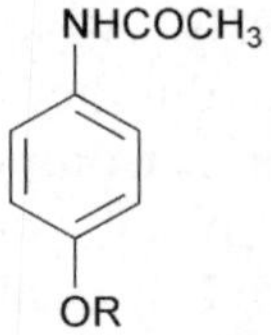   5.20 R = H
5.21 R = Et

Paracetamol **5.20** is widely used as a milder analgesic as it causes less gastric disturbance, although an overdose produces serious liver damage. The ethyl ether, phenacetin **5.21**, is a pro-drug for paracetamol. The presence of the ethyl group led to better absorption. However, it also influences the metabolism and this leads to hepatotoxicity. Hence phenacetin is no longer used.

In 1884, the pyrazol-5-one **5.22** and antipyrine **5.23** were found to have analgesic and anti-pyretic activity and formed a starting point for the synthesis of a series of analogues including aminopyrine **5.24**. Phenylbutazone **5.27** is prepared from butyl diethylmalonate **5.25** and diphenylhydrazine **5.26**.

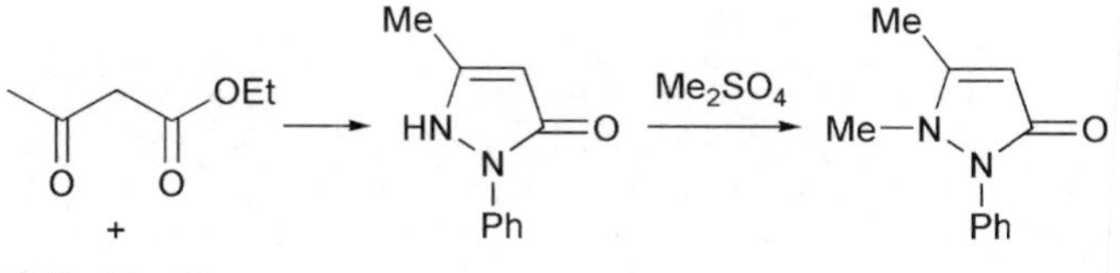

5.22   5.23 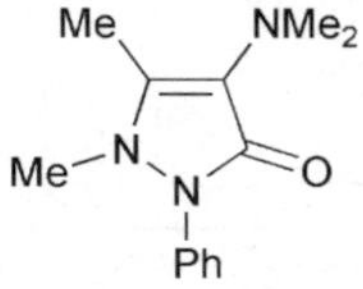   5.24

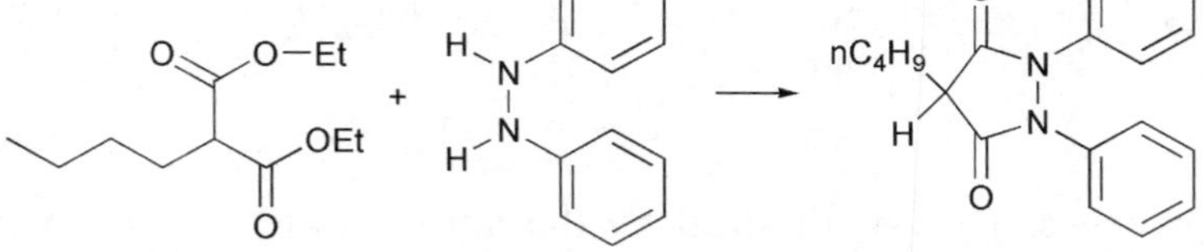

5.25   5.26   5.27

5.6   THE DEVELOPMENT OF IBUPROFEN

Although aspirin is an effective analgesic and is well established for the treatment of rheumatic inflammation of joints, it does have adverse side effects. Consequently, there have been efforts made to develop better non-steroidal anti-inflammatory agents. One of these is ibuprofen (Nurofen®) **5.33**. The development of this compound started in the mid-1950s. A bioassay was developed involving the induction of inflammation on the skin of a test animal, a guinea pig using ultraviolet radiation. Compounds were screened against this. The lead compound was a phenoxyacetic acid derivative **5.28**, which had initially been prepared in connection with agrochemical studies on selective herbicides. The ether oxygen was found to be unnecessary and good activity was shown to be present in para-alkylphenylacetic acids, *e.g.* **5.29**. Clinical trials were undertaken with ibufenac **5.30**. However, this compound was not well tolerated and it was replaced with ibuprofen **5.33**. This compound has a chiral centre in the propionic acid side chain. Only one isomer, the S isomer **5.34**, is active although the R isomer can be isomerized by man to the S isomer *via* the co-enzyme A ester. Ibuprofen was released for sale as a prescription medicine in 1969, and in 1983 it was given approval for 'over-the-counter' sales.

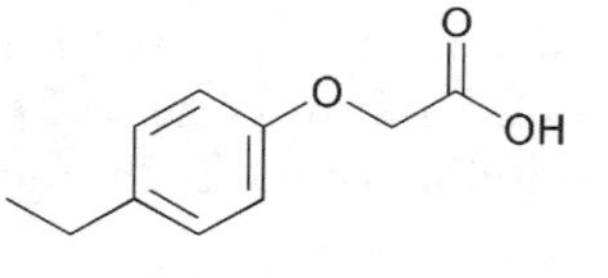

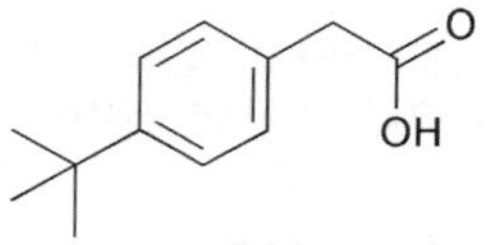

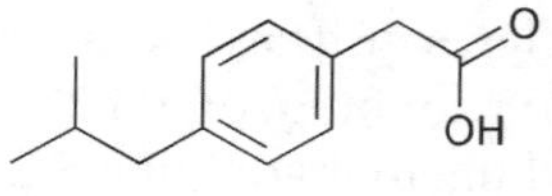

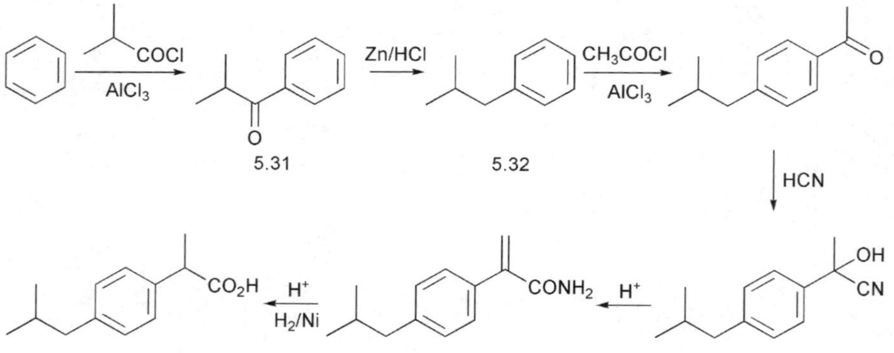

Ibuprofen can be synthesized by the above route. In this synthesis, note how an acyl group is used in the preparation of **5.31** from benzene to prevent an isomerization in the Friedel Crafts reaction. If a Friedel Crafts alkylation had been carried out with l-chloro-2-methylpropane to prepare **5.32** directly, there would have been some isomerization of the primary carbonium ion to a tertiary carbonium ion and a mixture would be obtained. Reduction has to take place before the second acyl group is introduced in order to obtain the correct orientation of the substituents. It is possible to see how structural modifications can be made in this synthesis by, for example, using a different acylating group.

Other compounds with useful NSAID activity include fenbufen **5.35**, indomethacin **5.36** and piroxicam **5.41**. The synthesis of piroxicam from saccharin **5.37** reveals an interesting ring enlargement reaction involving the ring cleavage of **5.38–5.39** and then re-cyclization by a Claisen condensation to form **5.40**. Diclofenac was introduced in Japan in 1974 and in other countries through the 1980s. It has powerful anti-inflammatory, analgesic and anti-pyretic properties. It appears to act at several steps in the metabolism of arachidonic acid including both the cyclo-oxygenase and lipo-oxygenase pathways.

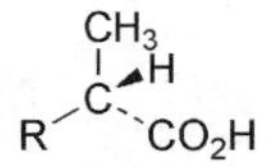

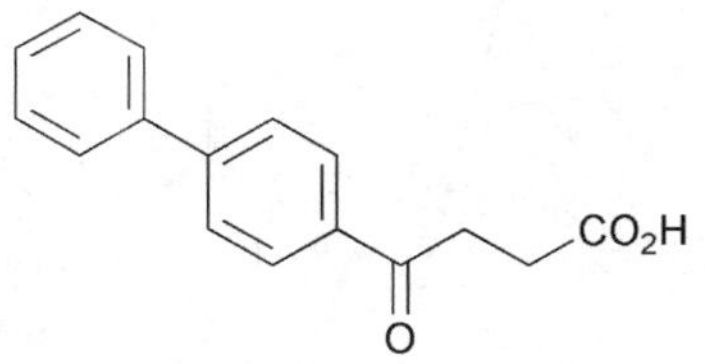

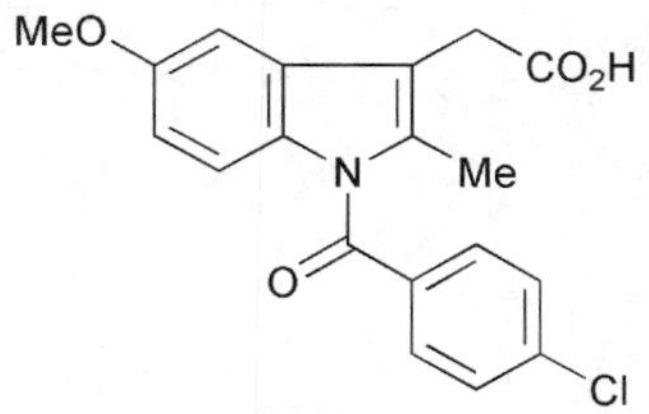

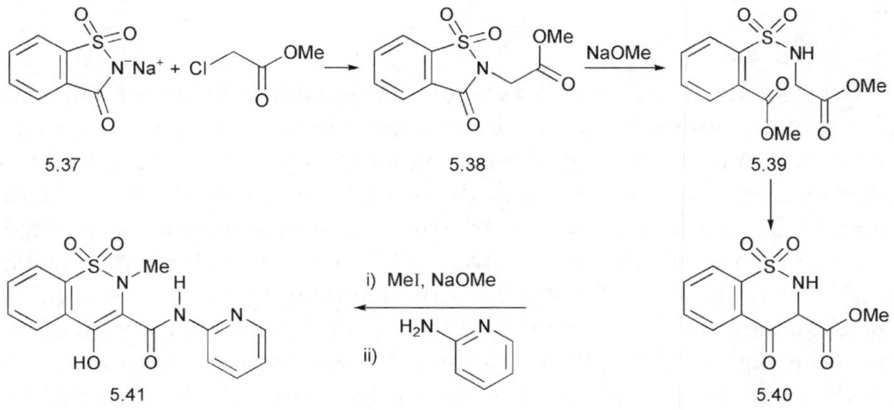

5.7   THE MECHANISM OF ACTION OF ASPIRIN

It was not until 1971 that the mechanism of action of aspirin was established by Vane. There is a widespread family of hormones known as the prostaglandins that are formed from the unsaturated acid, arachidonic acid **5.42**. This biosynthesis involves the addition of oxygen by an enzyme known as cyclo-oxygenase to give an endoperoxide **5.43**. Aspirin binds covalently to a serine residue within the enzyme and prevents the cyclization from taking place. Ibuprofen and piroxicam are reversible inhibitors while paracetamol affects oxygen uptake by the enzyme system.

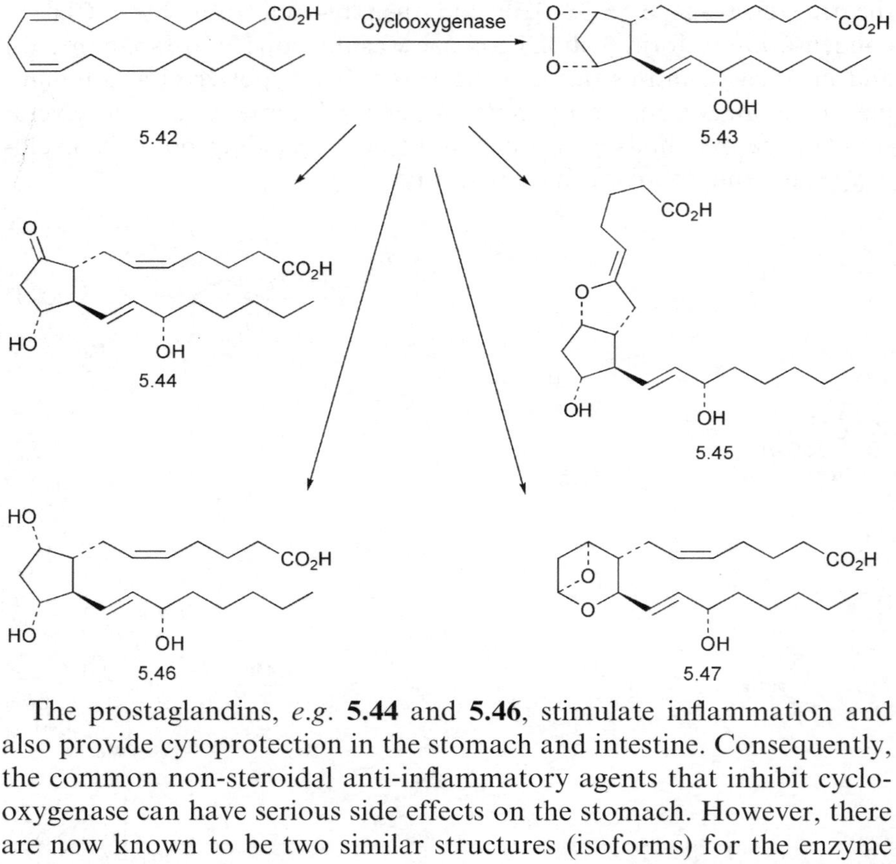

The prostaglandins, *e.g.* **5.44** and **5.46**, stimulate inflammation and also provide cytoprotection in the stomach and intestine. Consequently, the common non-steroidal anti-inflammatory agents that inhibit cyclo-oxygenase can have serious side effects on the stomach. However, there are now known to be two similar structures (isoforms) for the enzyme cyclo-oxygenase, COX-1 and COX-2. COX-1 is active all the time while COX-2 is transiently activated by pro-inflammatory substances such as the cytokines. Its activity is reduced by the anti-inflammatory steroids. It has been suggested that the constitutive COX-1 among other functions, produces the prostaglandins that protect the gastro-intestinal tract while the inducible COX-2 mediates inflammation. The difference in function provides an opportunity to separate toxicity from efficacy in the NSAIDs. A number of selective COX-2 inhibitors such as celecoxib (Celebrex®) **5.49** and rofecoxib (Vioxx®) have been synthesized, for the treatment of inflammation. Celecoxib **5.49** is prepared from 4-methyl-acetophenone **5.48**. However, cardiovascular side effects of these compounds have been reported recently and Vioxx has been withdrawn.

The cyclic endoperoxide **5.43** that is formed by cyclo-oxygenase is not only the precursor of the prostaglandins but also of the prostacyclins (*e.g.* PGI$_2$) **5.45** and the thromboxanes (*e.g.* TxA$_2$) **5.47**. The prostacyclins are produced by the endothelium of blood vessels and have a vasodilating and platelet anti-aggregation effect while thromboxane A$_2$ is released from blood platelets and has vasoconstricting and strong platelet aggregation properties. In a healthy circulating system there is a balance between these two contrasting hormones. Aspirin, by inhibiting cyclo-oxygenase, irreversibly inhibits the formation of both the prostacyclins **5.45** and the thromboxanes **5.47**. Prostacyclin biosynthesis recovers more rapidly than thromboxane biosynthesis as new cyclo-oxygenase is formed in the endothelial blood vessels. Hence the overall effect is a platelet anti-aggregation effect. This forms the basis of the protective effect of low doses of aspirin against stroke.

5.8 MEDICINAL USES OF PROSTAGLANDINS

Despite their widespread occurrence and a very considerable research effort, the use of prostaglandins or prostaglandin analogues themselves in medicine, has been limited. Among the reasons for this are the length of syntheses, problems in achieving selectivity and their rapid metabolism. Some applications include the use of prostaglandins E$_2$ and F$_2$ (dinoprostone **5.44** and dinoprostol **5.46**) in the induction of childbirth, misoprostol **5.50** in the treatment of peptic ulcers and latanoprost **5.51** in the treatment of glaucoma.

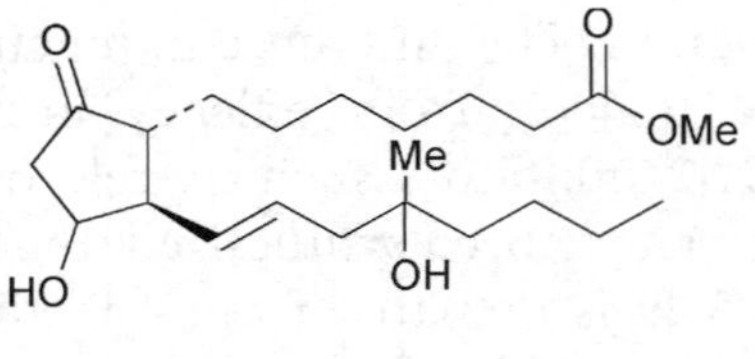

5.50

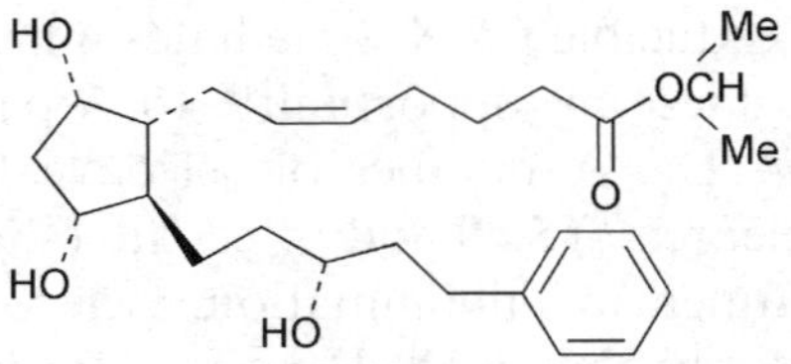

5.51

5.9 THE STEROLS AND STEROID HORMONES

The steroids are a widespread group of biologically active substances. They are characterized by a common carbon skeleton and they differ from each other by the nature of the substituents on this ring system. They are biosynthesized by the same general pathway.

Lanosterol **5.52** is formed by the cyclization of squalene epoxide in the first of the steps that lead to the sterols and steroid hormones. It is a constituent of wool fat and its esters are components of lanolin cream. Cholesterol **5.53** is an important membrane component and, in the course of being transported in the circulatory system, it forms complexes with protein that can be deposited and block the circulatory system. Low-density lipoprotein (LDL) is particularly serious in this context.

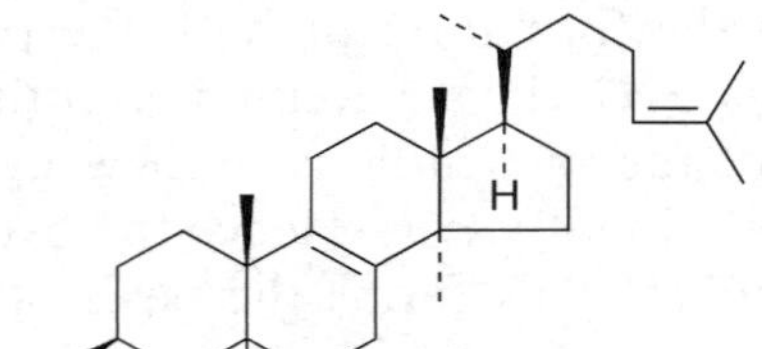

5.52

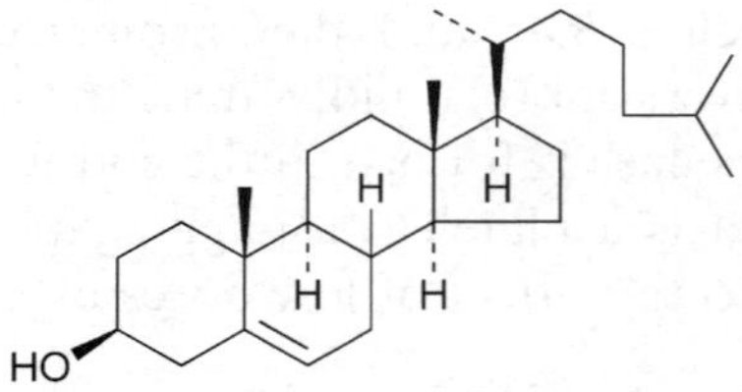

5.53

Vitamin D_3 **5.54** is a metabolite of 7,8-dehydrocholesterol. The main function of 1α, 25-dihydroxyvitamin D_3 (calcitriol) **5.55** in humans is to control gastrointestinal calcium and phosphate ion absorption and to promote mineralization of bones. A deficiency of this vitamin leads to rickets in children and to a softening of the bones in adults. It is also involved in cell differentiation processes. The cholic acids (the bile acids), *e.g.* **5.56** are involved in the emulsification and transport of fats.

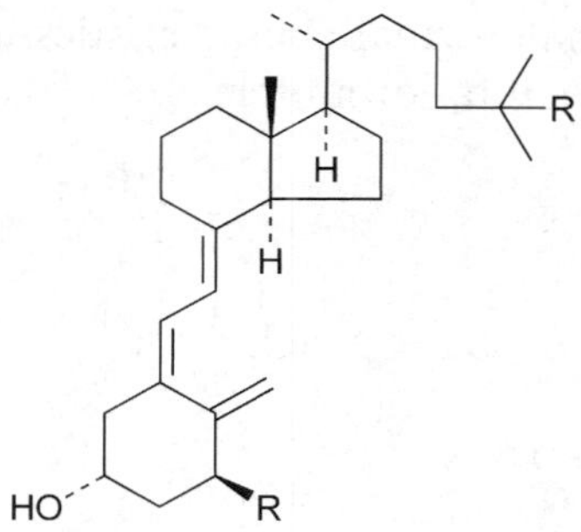

5.54 R = H
5.55 R = OH

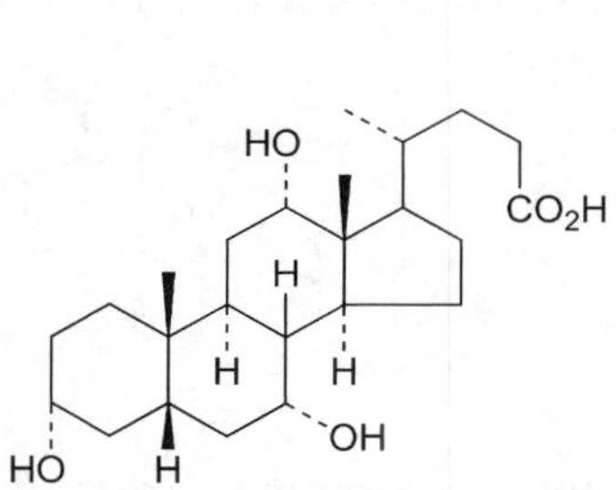

5.56

Whereas cholesterol and the bile acids are produced in gram quantities in the body, the hormonal steroids are found in milligram quantities. The steroidal cortical hormones, *e.g.* cortisol **5.57**, are produced by the adrenal cortex and are involved in the regulation of carbohydrate, lipid and protein metabolism (glucocorticoid effects) and in the regulation of electrolyte balance (mineral corticoid effects). They also have immuno-suppressive activity.

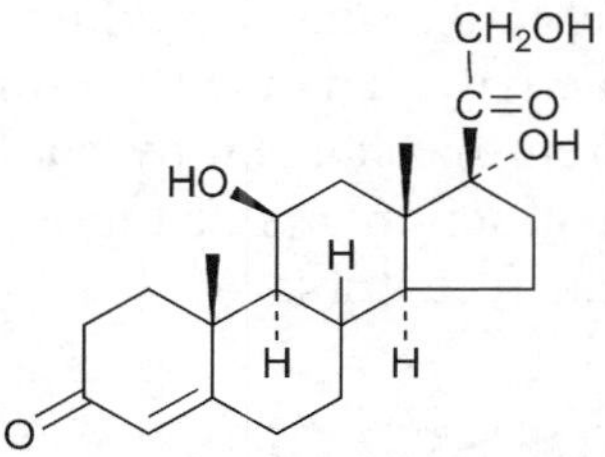

5.57

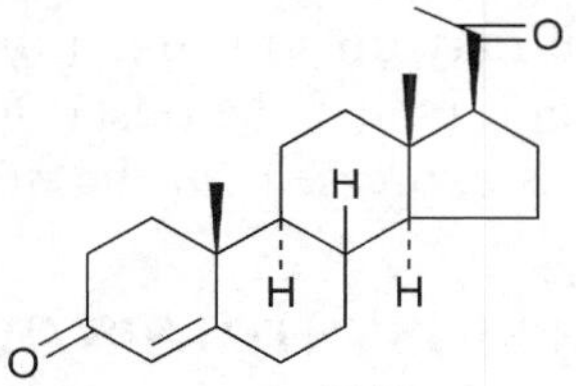

5.58

The progestogens, *e.g.* progesterone **5.58** and the estrogens, *e.g.* estradiol **5.60** are female sex hormones. The estrogens are formed mainly in the ovary. Each ovary contains numerous follicles in which the egg develops. When the egg reaches maturity and ovulation occurs, the follicle ruptures. The corpus luteum is formed at the site of the ruptured follicle and produces progesterone. This targets the lining of the uterus and if fertilization has occurred, it facilitates implantation and the maintenance of pregnancy. During pregnancy, progesterone prevents the ripening of further follicles and it stimulates the development of the lactic glands. Various pregnanes also act on the brain and can affect mood. The estrogens, *e.g.* estradiol **5.60**, produce the female sexual characteristics. They are responsible for the proliferation of uterine mucosa and their production affects the female menstrual cycle. The androgens, *e.g.* testosterone **5.59**, are produced by the testes and are responsible for the development of male sexual characteristics and are anabolic steroids, stimulating the formation of muscle.

5.59   5.60

The steroid hormones are hydrophobic and are transported in the body from their site of synthesis to their target cell by carrier proteins. These complexes dissociate on reaching the target cell. The steroid hormones then exert their biological effect by being transported into the cell and by binding to the nuclear receptor. When the steroid binds to this receptor protein, a portion of the protein known as the 'chaperone' or 'heat-shock protein' dissociates. The steroid-receptor protein complex then dimerizes and binds to a specific DNA sequence. This leads to activation of the DNA and, *via* mRNA, to the expression of biological activity. Ring A of many of the steroids is similar and there is a model for the biological activity based on ring A binding and ring D expressing the particular biological activity. The neurosteroids that act in the brain may do so using a different mechanism by binding to cell-surface receptors and the onset of their action is much more rapid than that which is expressed *via* the nuclear receptors.

5.10   THE BIOSYNTHESIS OF THE STEROIDS

The biosynthesis of the steroids provides a basis for understanding many aspects of their medicinal chemistry. The steroids belong to a much larger family of natural products known as the isoprenoids or terpenoids. The carbon skeleta of these compounds are assembled from C_5 isoprene units. In the case of the steroids six of these units are involved. Once all six have been assembled *via* two C_{15} units and the cyclization has taken place, there is a stepwise degradation of the C_{30} carbon skeleton to form the individual compounds. The key steps in the biosynthesis of cholesterol from acetate units involve hydroxymethylglutaryl co-enzyme A **5.61**, mevalonic acid **5.62**, farnesyl pyrophosphate **5.63** and squalene epoxide **5.64**. The side chain of cholesterol **5.53** is cleaved to give pregnenolone, which is oxidized to progesterone **5.58**. Progesterone is the precursor of the cortical steroids, *e.g.* **5.57** and of the androgens such as testosterone **5.59**. In turn testosterone is converted by an enzyme system known as aromatase, to **5.60**.

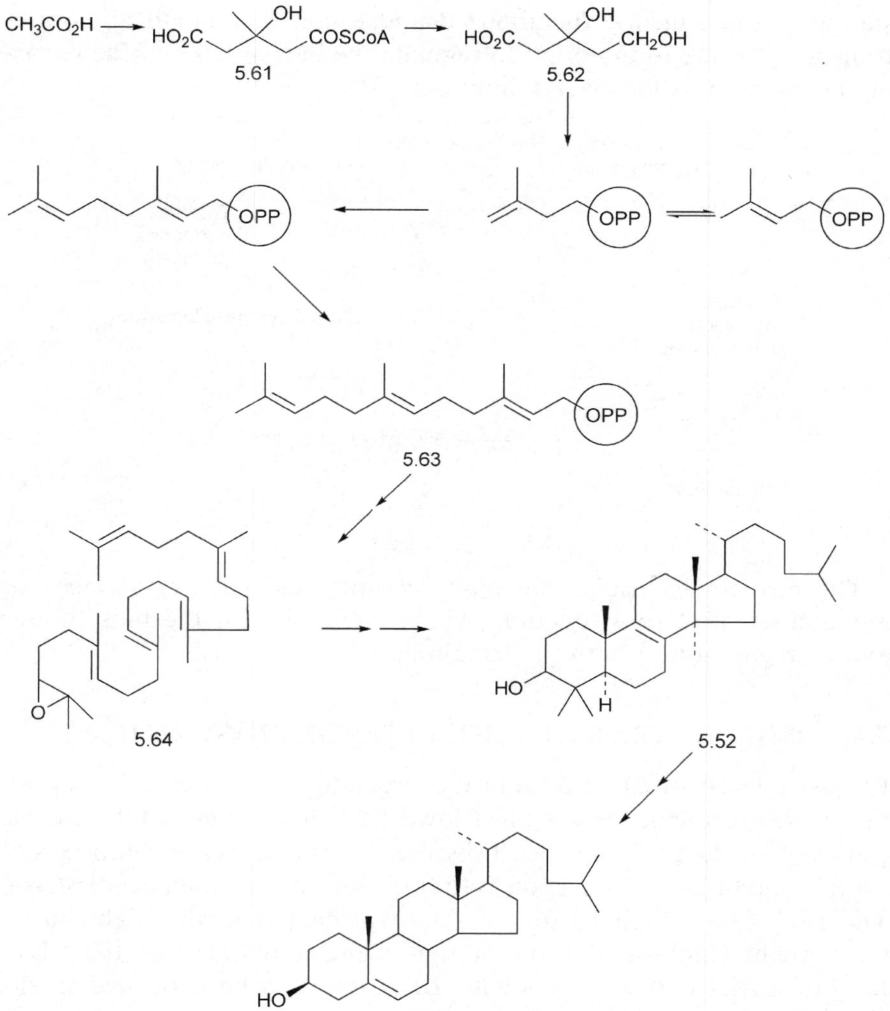

The biosynthesis of the hormonal steroids is regulated by a steroid cycle (see **5.65**) in which a number of peptide hormones play a role. The biosynthesis is initiated by the hypothalamus in the brain. This is the source of peptide releasing hormones that stimulate the pituitary gland into producing the peptide trophic hormones, for example, ad-renocorticotrophic hormone (ACTH), and follicle stimulating hormone (FSH). The trophic hormones have the steroid synthesizing organs, such as the adrenal glands and the ovaries, as their target. The steroid hormones that are formed are then transported to their site of action where they produce a response. Although the major portion of the steroid is subsequently metabolized and excreted, a small amount is transported back to the brain to regulate the biosynthesis of the releasing hormone by a feedback inhibition.

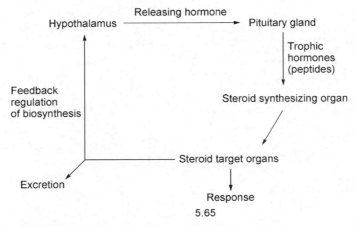

5.65

The medicinal chemist can make compounds that are agonists or antagonists of steroid function, which interfere with the biosynthesis and its regulation or with the metabolism of the steroids.

5.11 THE CONTROL OF CHOLESTEROL BIOSYNTHESIS

Excessive levels of cholesterol in the circulatory system and its deposition with a protein, the so-called lowdensity lipoprotein LDL, are the causes of stroke and some heart disease. Lipoproteins are complexes of protein and lipids and their role is to transport fats including cholesterol. The LDL has a higher ratio of fat to protein than the high-density lipoprotein. Cholesterol forms a significant component of the LDL. Because of its higher fat content, LDL tends to be deposited in the circulatory system and can form blockages.

The control of cholesterol levels can be achieved partly by diet and partly by the use of cholesterol biosynthesis inhibitors. The statins, such as compactin **5.66**, mevinolin (lovastatin) **5.67** and simvastatin (Zocor®) **5.68**, are examples of drugs that are of major importance. These compounds act as inhibitors of hydroxymethylglutaryl co-enzyme A **5.61** reductase and hence of the biosynthesis of mevalonic acid **5.62** and the C_5 isopentenyl pyrophosphate building block of sterol biosynthesis. All of these drugs contain a β-hydroxy-δ-lactone reminiscent of the lactone of mevalonic acid. In atorvastin (Lipitor®), the lactone is hydrolysed and the drug is presented as the calcium salt of the hydroxy acid. Other agents that prevent cholesterol deposition include the anion exchange resins cholestyramine, and niacin (nicotinic acid) **5.69**. The fibrates, clofibrate **5.70** and gemfibrozil **5.71**, also lower serum lipoprotein concentration.

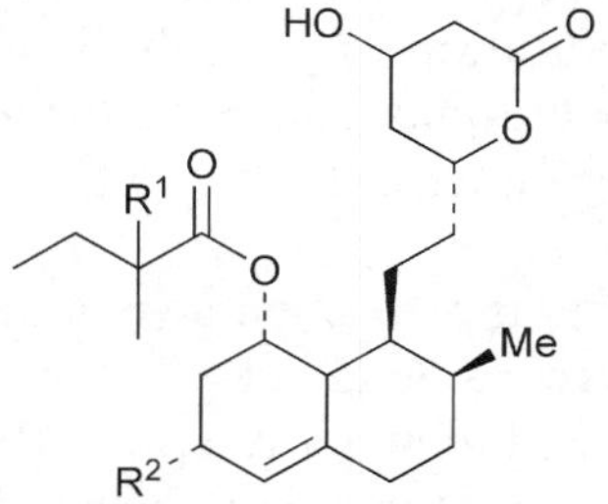

5.66 $R^1 = R^2 = H$ mevastatin, compactin
5.67 $R^1 = H$, $R^2 = Me$ lovastatin, mevinolin
5.68 $R^1 = R^2 = Me$ simvastatin

The gall bladder produces the bile acids from cholesterol. The bile acids have a concave structure with a lipophilic and a hydrophilic face with the hydroxyl groups on the α (underside) of the molecule. Gallstones that form in the gall bladder are mainly cholesterol deposits arising from the incomplete conversion of cholesterol into the bile acids. Gallstones may be removed by surgery or disintegrated ultrasonically. However, small gallstones can be dissolved by the action of the less common bile acids, chenodeoxycholic acid **5.72** and ursodeoxycholic acid **5.73**.

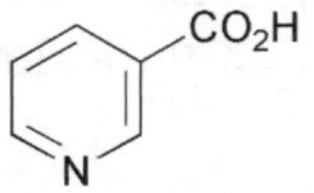

5.69

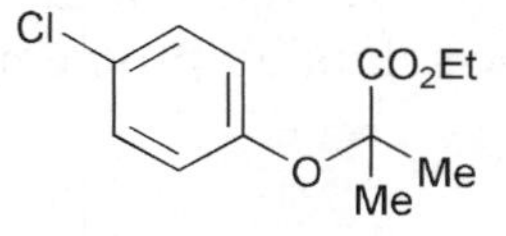

5.70

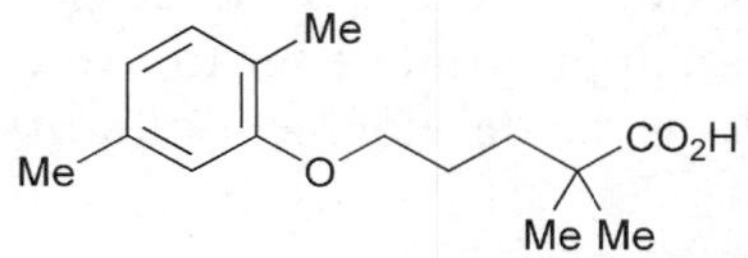

5.71

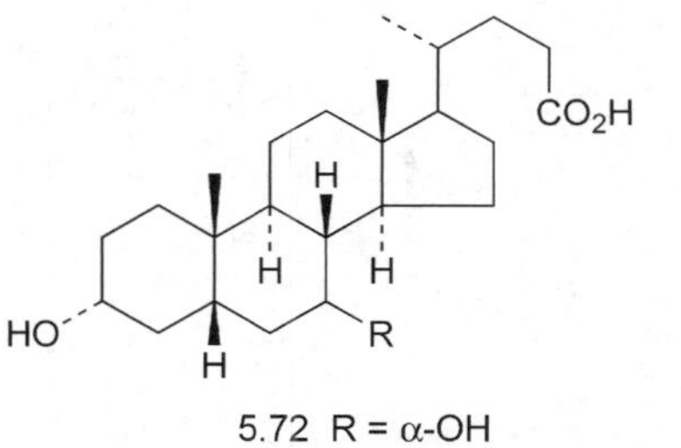

5.72 R = α-OH
5.73 R = β-OH

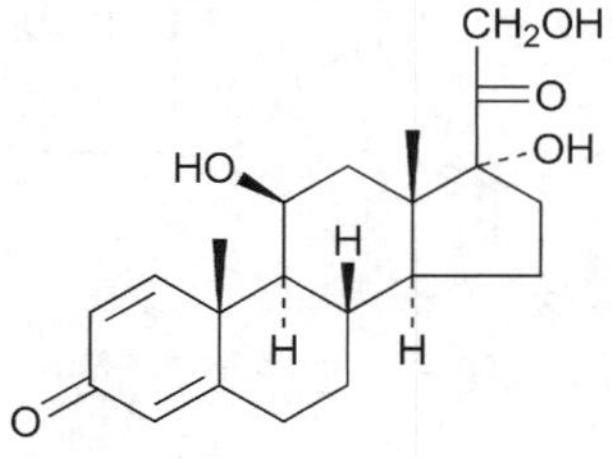

5.74

5.12   THE STEROIDAL ANTI-INFLAMMATORY AGENTS

The structures of the steroid hormones that are produced by the adrenal glands were established and a number were synthesized from more readily available steroids in the 1930s and 1940s. The discovery in 1949 of the anti-inflammatory action of one of these, cortisone, revolutionalized the treatment of rheumatoid arthritis. This stimulated a great deal of research in the 1950s and 1960s leading to useful steroidal anti-inflammatory agents.

Several mechanisms may be involved. The first is an immunosuppressive effect and the second is an indirect inhibition of phospholipase A2. This enzyme mediates the release of arachidonic acid **5.42** from lipid reservoirs. Arachidonic acid is the starting material for the biosynthesis of the pro-inflammatory prostaglandins.

Steroid therapy can produce significant improvements in many inflammatory diseases but there can be adverse side effects. The mineralcorticoid activity of these compounds can lead to water retention. Hence, there have been structural variations to increase the gluco-corticoid anti-inflammatory activity and to diminish the mineral-corticoid activity and thus to reduce the amount of steroid required for treatment. Prednisolone **5.74** is the most widely used steroid for the treatment of inflammatory and allergic diseases. Betamethasone **5.75**, dexamethasone **5.76**, triamcinolone **5.77** and synalar **5.78** are other examples. Where it is necessary to transport the drug across fat barriers, for example, in topical creams that are used to treat local inflammation, the steroid may be converted to more lipid-soluble C-21 esters. Where a water-soluble form is required the C-21 hydroxyl group is converted to a hemi-succinate or monophosphate and the steroid is dispensed as the sodium salt. The use of an appropriate ester to overcome solubility problems is an important step in the development of a drug.

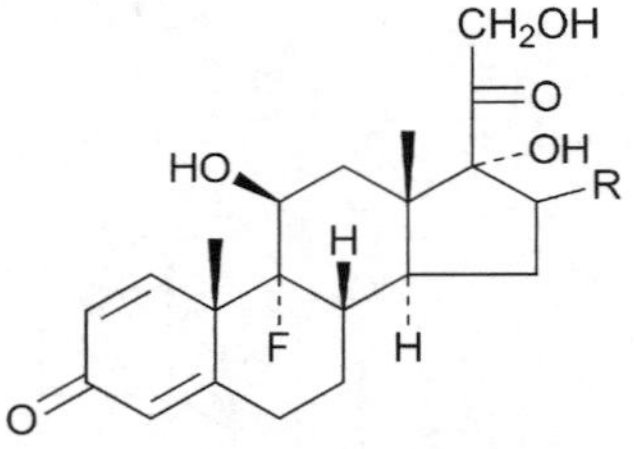

5.75 R = β-Me
5.76 R = α-Me

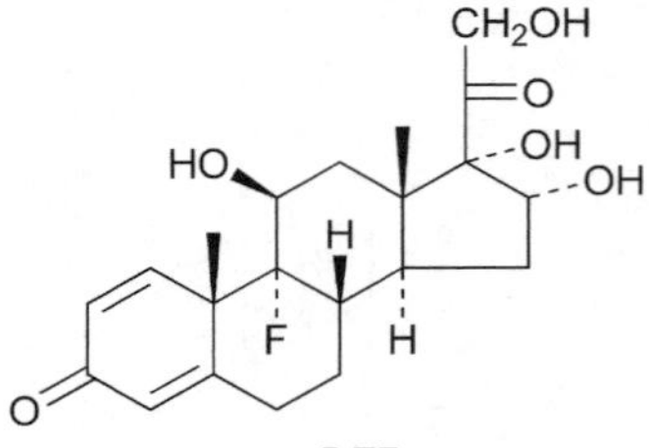

5.77

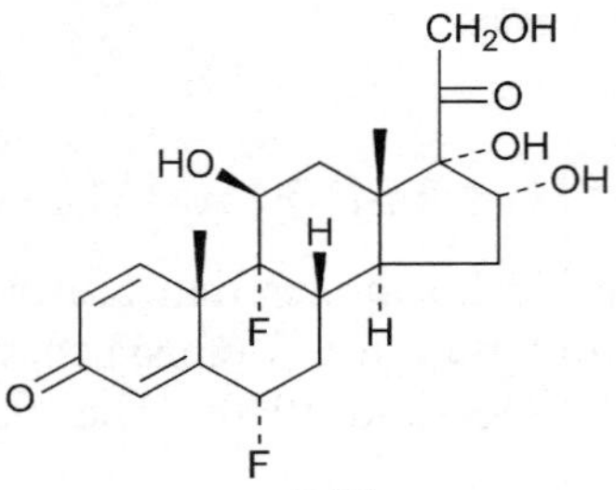

5.78

5.13 THE STEROIDAL ORAL CONTRACEPTIVES

Hormonal contraception has brought about significant social change. It was known that once a female was pregnant, further pregnancies did not occur. In 1921 it was shown that rabbits could be made infertile by transplanting the ovaries or corpus luteum from a pregnant rabbit and placing it under the skin of the test rabbit. The conclusion was drawn that compounds were being produced that inhibited ovulation. In the late 1920s, two groups of compounds, the estrogens and the progestogens, were isolated from the urine of pregnant animals. Their structures were established in the 1930s. Soon after these compounds were isolated, it was demonstrated in animal experiments that injection of these compounds inhibited ovulation and created a state of pseudo-pregnancy. However, they were rapidly metabolized and hence they had a relatively weak effect when administered orally. In 1938 it was found that 17a-ethynylestradiol **5.79** was effective by an oral route; 17a-ethynyltestosterone **5.81** was also effective. In 1951 Pincus and others began to develop the oral contraceptive. The first commercial product, Enovid®, was available in 1960. The typical steroids that are used include **5.79–5.84**.

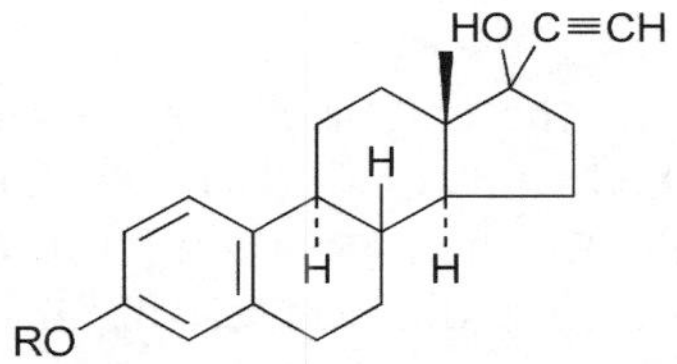

5.79 R = H
5.80 R = Me

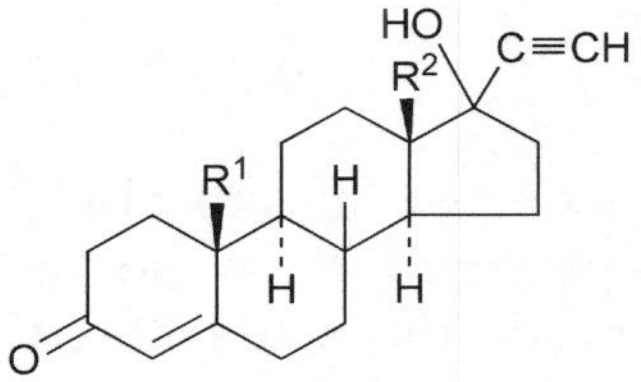

5.81 $R^1 = R^2$ = Me
5.82 R^1 = H, R^2 = Me
5.83 R^1 = H, R^2 = Et

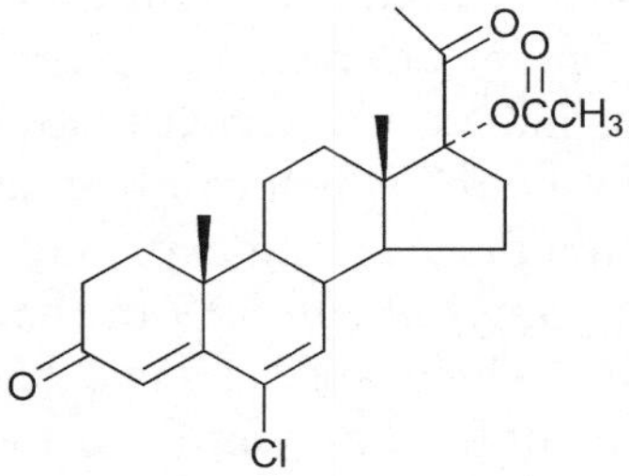

5.84

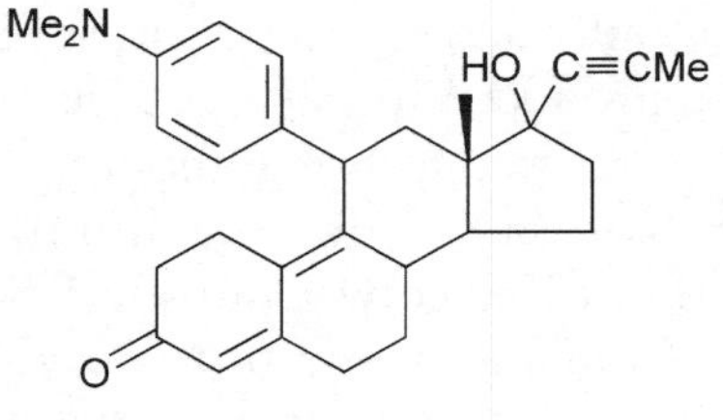

5.85

These compounds function by acting in the regulatory feedback loop of the steroid cycle and switch off the natural production of steroid hormones. Some also affect the vaginal mucosa and thus sperm motility.

A completely different effect is brought about by the progestogen antagonist, RU 486 (mifepristone) **5.85**. This compound binds to the progesterone receptor but it does not facilitate implantation and it prevents the development of a pregnancy, the so-called 'morning after pill'.

The development of a number of cancers are hormonally dependent. Compounds that affect steps in their biosynthesis, and which are antagonists at their receptors, are discussed in the chapter on cancer chemotherapy.

5.14 THE ROLE OF NITRIC OXIDE

Nitric oxide is a gas that is produced in the body by the action of NO synthase on the amino acid, arginine **5.86**. The terminal guanidine unit of this amino acid is oxidized to generate nitric oxide and another amino acid, citrulline **5.87** that contains a urea. Nitric oxide is a signalling molecule that, when it is produced in one cell, permeates the adjacent cells. There it activates the enzyme that forms cyclic guanosine monophosphate (GMP) from guanosine triphosphate. Cyclic GMP behaves as a second messenger and activates further enzymes.

          5.86   5.87

Nitric oxide was identified in 1987 as the endothelial-relaxing factor, which is produced in the endothelial cell walls of the arteries. The nitric oxide spreads through the cell walls to the underlying muscle cells relaxing their constriction and leading to a dilation of the arteries. In this way it controls blood pressure. It also causes a relaxation of the gastrointestinal muscle and this affects stomach motility. The macrophages that are part of the immune response to infection, also produce nitric oxide to kill bacteria. In the case of a severe infection enough nitric oxide may be produced to induce sufficient vasodilation to seriously lower the blood pressure and bring about a 'septic shock', which can be fatal.

The discovery of the role of nitric oxide provided an explanation for the action of glyceryl trinitrate **5.88** and amyl nitrite **5.89** in the treatment of angina. These nitrovasodilators, which are a source of nitric oxide, are used to treat cardiovascular problems arising from the narrowing of arteries

          5.88   5.89

CHAPTER 6

Anti-Infective Agents

6.1 AIMS

The aim of this chapter is to describe the development of compounds with a selective action against invasive organisms. By the end of this chapter, you should be aware of

- the role of the sulfonamides and the penicillins and their mode of action in the control of bacterial infections;
- the development of anti-viral agents, which interfere with nucleic acid biosynthesis and viral recognition;
- the development of azole anti-fungal agents and their mode of action in the inhibition of ergosterol biosynthesis; and
- the role of medicinal chemistry in the control of malaria.

6.2 INTRODUCTION

The control of infectious diseases associated with micro-organisms has been one of the major triumphs of medicinal chemistry. However, the development of resistant organisms now presents one of the current challenges. In the mid-nineteenth century, it was realized that plant diseases such as potato blight were caused by fungi and that the spoilage of beer and wine was caused by bacteria. This formed the basis for the investigations of microbial diseases of man. The work of Pasteur in the 1860s in identifying bacteria as disease-producing organisms and of Lister who introduced phenol as an antiseptic, was of major importance. In 1876, Koch identified the anthrax bacillus as the causative organism for anthrax, and in 1882 the tubercule bacillus as the cause of tuberculosis. The Gram stains were introduced in 1884 to distinguish between different groups of bacteria. This selective uptake of dyestuffs by bacteria provided the basis for the exploration by Ehrlich of the use of dyestuffs as anti-microbial agents. Ehrlich introduced the term 'chemotherapy' to cover the use of drugs to injure an invading organism without injury to the host. The development of many dyestuffs based on the coal tar industry during the late nineteenth and early twentieth centuries, provided ample lead compounds.

6.3 BACTERIAL DISEASES

Some clinically important bacteria include the following. *Staphylococcus aureus* is a Gram-positive organism, which causes skin and soft tissue infection and can cause septicaemia. It forms the yellow pus around an infected wound. Strains that are resistant to the antibiotic methacillin (methicillin resistant *Staphylococcus aureus* = MRSA) are a serious hospital problem. *Streptococcus* species are Gram-positive organisms that are responsible for sore throats and other upper respiratory tract infections.

Clostridium species are another genus of Gram positive, mostly anaerobic organisms that include some serious intestinal pathogens. The bacterium *Cl. botulinum* growing on meat, produces the toxins responsible for botulism, *Cl. tetanii* is the cause of tetanus when it infects a deep wound, while *Cl. deficile* is the source of a number of hospital infections. *Escherichia coli* and *Salmonella* species are Gram-negative organisms that are involved in gastro–intestinal tract infections and in typhoid. Other Gram-negative organisms such as *Enterobacter*, *Klebsiella* and *Proteus* species are also responsible for respiratory infections, while *Haemophilus influenzae* is a Gram-negative organism associated with chest and ear infections and with bacterial meningitis. Syphilis is a sexually transmitted disease caused by a bacterium, *Treponima pallidum*. Although an organism may typically be associated with a disease, some strains of the organism may be more virulent than others, a feature found with *E. coli*. The strain *E. coli* 0157, sometimes known as the verocytotoxin-producing strain, can be fatal.

As part of a bacterial infection, the bacteria produce toxic proteins, antigens, to which certain cells in the mammalian body respond by producing a complimentary protein, an antibody. The antibody combines with the antigen to form an innocuous product. The effects of this immune response include the inflammation and fever associated with a viral or a bacterial infection.

The generalized structure **6.1** of a bacterium illustrates the differences in cellular structure compared to the mammalian cell. These differences provide the basis for the selective toxicity that underpins the biological activity of many anti-bacterial agents. Bacteria are prokaryotes, *i.e.*, they have no nucleus-containing DNA. The chromosomes containing the nucleic acid are within the cellular structure. This contains the ribosome where protein synthesis occurs initiated by the messenger RNA. A major difference is in the cell wall, which forms an important target. A large family of anti-microbial agents inhibit bacterial cell wall biosynthesis and affect the permeability of the plasma membranes. There are several other targets for anti-microbial agents. These include the inhibition of essential biosynthetic pathways and in preventing the formation and the function of co-enzymes particularly those that are found in bacteria and not in man. Nucleic acid biosynthesis and protein biosynthesis in the ribosomes provide further specific targets.

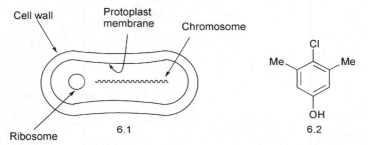

6.1   6.2

The bioassay involves seeding the agar in a Petri dish with a test organism such as *Staphylococcus aureus*. A small sample of the putative anti-bacterial agent is placed in the centre of the plate. The plate is then incubated at 37°C for 24 h. There will be a zone of inhibition, the diameter of which is proportional to the concentration and anti-microbial activity of the substance being tested.

6.4   ANTISEPTICS

Since 1865 when Lister first used phenol as an antiseptic, a number of simple compounds have been developed for use as local anti-bacterial agents. These include chloroxylenol **6.2** (dettol), thymol **6.3** and hexachlorophene **6.5**, which is prepared from trichlorophenol (TCP) **6.4** and formaldehyde.

6.3   6.4   6.5

Quaternary ammonium salts represent another class of antiseptic agents, which include cetrimide **6.6** and cetylpyridinium bromide **6.7**.

Chlorhexidine **6.8** (hibitane) is made from hexane-1,6-diamine and sodium dicyanamide. Many of these compounds exert their action by surface activity effects, which damage the bacterial membranes and lead to rupture of the bacterial cell.

Me(CH$_2$)$_{13}$N$^+$Me$_3$Br$^-$ Me(CH$_2$)$_{15}$—N$^+$⟨⟩ Br$^-$

     6.6   6.7

H$_2$N(CH$_2$)$_6$NH$_2$ + 2NaN(CN)$_2$ ⟶

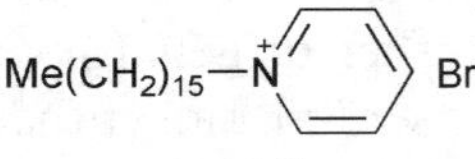

NC–NH–C(=NH)–NH–(CH$_2$)$_6$–NH–C(=NH)–NH–CN

Cl–⟨⟩–NH$_2$

Cl–⟨⟩–NH–C(=NH)–NH–C(=NH)–NH–(CH$_2$)$_6$–NH–C(=NH)–NH–C(=NH)–NH–⟨⟩–Cl

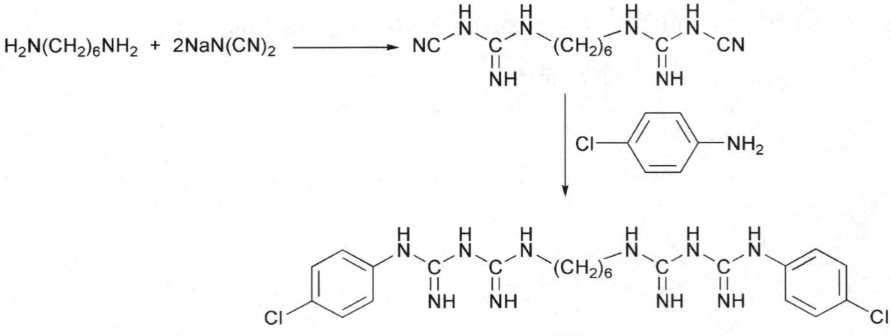

6.8

6.5 THE SULFONAMIDE ANTI-BACTERIAL AGENTS

Since some dyestuffs selectively stained micro-organisms, Ehrlich proposed that various pathogenic organisms might be controlled by the selective use of dyestuffs. Over the years, this led to many investigations and to some useful anti-bacterial substances including the acriflavine **6.9** anti-bacterial agents introduced in 1917.

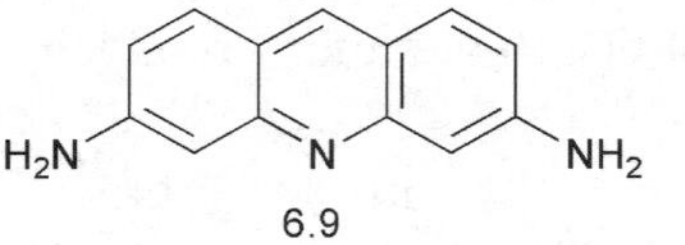

H$_2$N–⟨acridine⟩–NH$_2$ HO–⟨⟩(NH$_2$)–As=As–⟨⟩(OH)(NH$_2$)

     6.9 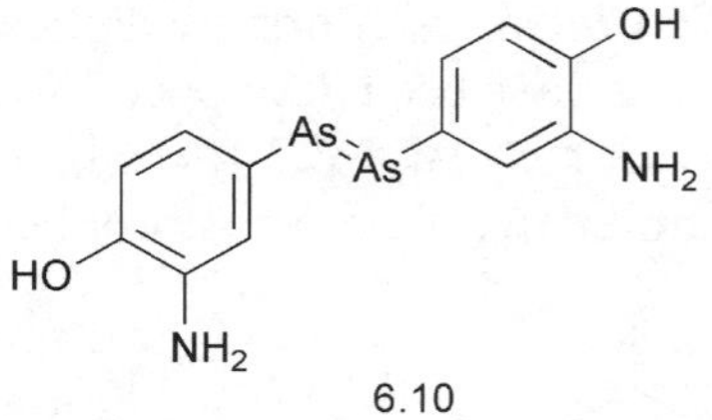   6.10

In 1909, the organoarsenic compound, salvarsan **6.11** was found to be active against trypanosomiasis (sleeping sickness) and syphilis. Salvarsan was derived from an earlier (1905) drug, *p*-aminophenylarsonic acid (atoxyl) and modelled on an azo-dye with the supposition that it would contain an arsenic:arsenic double bond in place of the nitrogen:nitrogen double bond. Formulations such as **6.10** were proposed. Subsequent work has shown that this is incorrect and that salvarsan is a polymer

6.11 in which the singly bonded arsenic atoms may lie in a ring. The current suggestion is that salvarsan is a mixture of cyclic species containing arsenic in three- and five-membered rings.

6.11   6.12   6.13

It was known that the sulfamyl ($-NHSO_2-$) group facilitated the binding of dyestuffs to wool protein. Hence, Domagk (1932) investigated the anti-bacterial activity of some dyestuffs including prontosil red **6.12**. Unlike today, the primary bioassay included whole animal tests and he was able to show that the compound was active against a *Streptococcus* infection in mice and a *Staphylococcus* infection in rabbits. Again unlike today, the compound was rapidly tested in humans where it was shown to control septicaemia in children including Domagk's daughter. It had a significant effect on infant mortality. It was soon realized (1935) that the active compound was a bio-transformation product, sulfanilamide **6.13**. However, although this was used as an anti-bacterial agent, it was accompanied by a side effect. The amino group was acetylated in man and the acetate crystallized out in the kidneys. A series of structural modifications were then made to overcome this problem.

6.14   6.15

The synthesis of sulfonamides is outlined in **6.14** and **6.15**. This synthesis is readily adapted to make a range of derivatives, *i.e.*, it fulfils one of the major criterion of a synthesis in medicinal chemistry, that of flexibility.

One of these, sulfapyridine (M and B 693) **6.16**, achieved fame because it was used to treat the wartime Prime Minister, Winston Churchill in 1943 when he had contracted pneumonia. Other sulfonamides that have been used include sulfathiazole **6.17**, sulfadiazine **6.18** and sulfamethoxazole **6.19**.

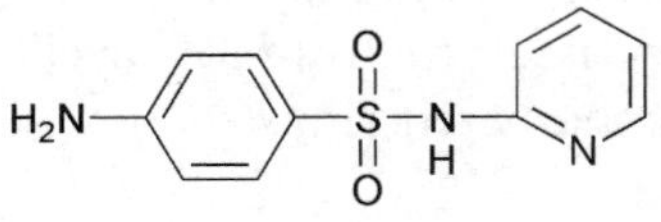

6.16

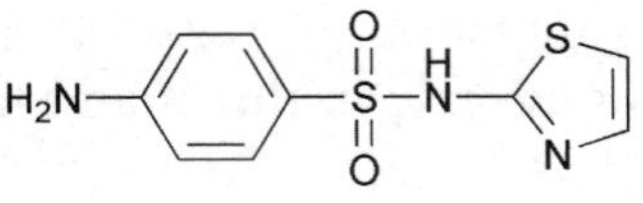

6.17

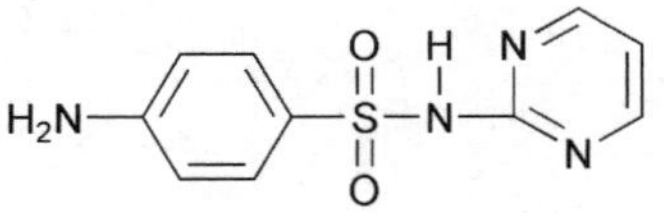

6.18

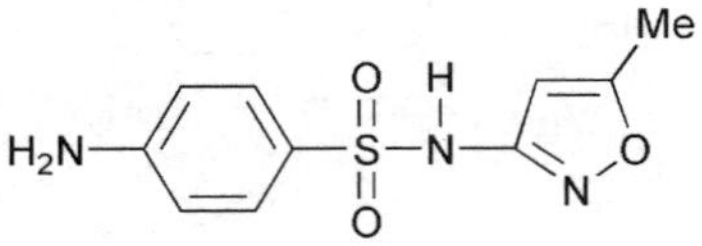

6.19

Structure:activity relationships showed a structural requirement for the *p*-aminophenylsulfonamide group in which the sulfonamide group should carry only one substituent. The N–H of the sulfonamide group is quite acidic. All of the heterocyclic rings attached to this grouping have an imino group and are probably hydrolysed enzymatically.

Insight into the mechanism of action of the sulfonamides came from an observation by Woods and Fildes in 1940 of the structural similarity between sulfanilamide **6.13** and *p*-aminobenzoic acid **6.20**.

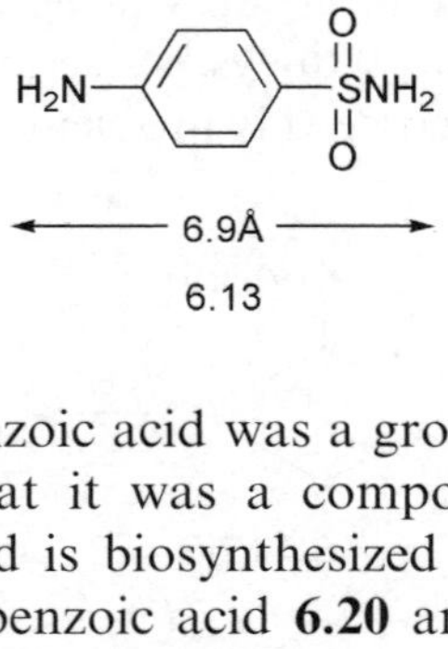

6.13

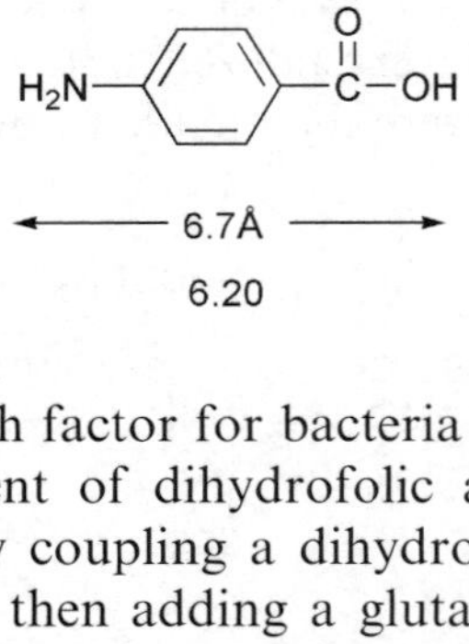

6.20

p-Aminobenzoic acid was a growth factor for bacteria and in 1946 it was found that it was a component of dihydrofolic acid **6.22**. Dihydrofolic acid is biosynthesized by coupling a dihydropteridine **6.21** with *p*-aminobenzoic acid **6.20** and then adding a glutamic acid unit. This biosynthesis is inhibited by sulfanilamide. Dihydrofolic acid is the precursor of tetrahydrofolic acid, which is the co-enzyme for a number of essential biosynthetic steps including the addition of the C_1 unit to uracil to form thymidine, a major component of the nucleic acids. In humans, folic acid is an essential dietary factor which is obtained from food. It is not biosynthesized by man. Consequently, the sulfonamides have no direct effect on folic acid in man but only on the production of folic acid in bacteria.

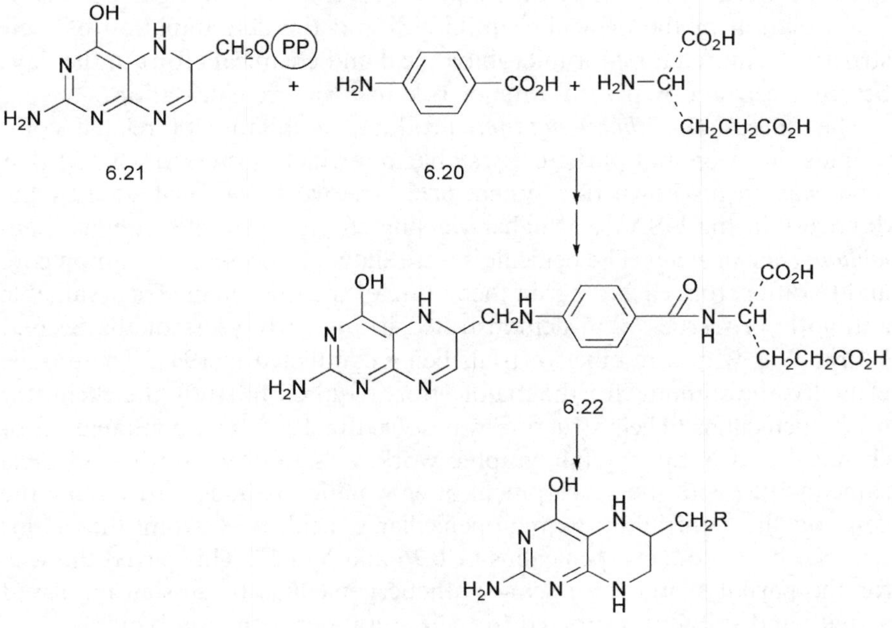

The conversion of dihydrofolic acid **6.22** to tetrahydrofolic acid **6.24** is inhibited by another anti-bacterial agent, trimethoprim **6.23**. A combination therapy of sulfamethoxazole **6.19** and trimethoprim (co-trimazole) has been particularly useful in combating resistance. If a bacterium develops resistance to sulfamethoxazole, it is killed by trimethoprim before it can replicate and so the resistance cannot be passed on. The enzyme dihydrofolate reductase has been crystallized and its X-ray crystal structure has been determined. The structure of the enzyme containing bound trimethoprim has also been determined.

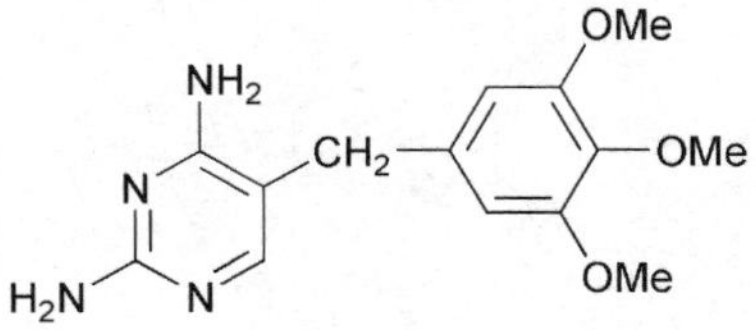

6.6 THE PENICILLINS

The observation by Fleming in 1928 of the antibiosis by the fungus *Penicillium notatum*, a chance contaminant of a culture of a *Staphylococcus* species proved to be one of the major discoveries of medicinal chemistry. The isolation of the penicillin antibiotics and the determination of their structure required a major microbiological and chemical effort in the days before spectroscopy played a major role in structure elucidation.

The fungus *Penicillium notatum* produced a mixture of related compounds that were not particularly stable and which possessed what at that time, was an unknown ring system **6.25**. The work was facilitated by the discovery in the USA of a higher-yielding strain of another fungus, *Penicillium chrysogenum*. The penicillins were shown to possess a common core and to differ from each other in the nature of the side chain. Their valuable antibiotic properties were demonstrated in the early years of the Second World War with a number of trials being conducted in 1941. These trials provided the stimulus for the major effort required to study the chemistry of the penicillins. Their structure was solved in 1945 by a combination of chemical and X-ray crystallographic work. A significant further advance came in 1959 with the development of enzymatic methods for isolating the core of the penicillins, 6-aminopenicillanic acid **6.28** from the major fermentation products, penicillins G **6.26** and V **6.27**. This paved the way for the partial synthesis of semi-synthetic penicillins possessing improved activity and stability compared to the natural-occurring antibiotics.

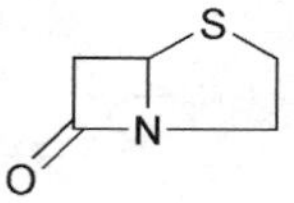

6.25

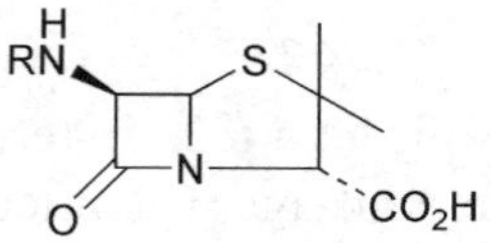

6.26 R = phenyl—$CH_2C(=O)$—

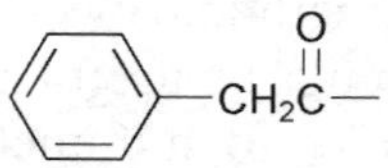

6.27 R = phenyl—O—$CH_2 \cdot C(=O)$—

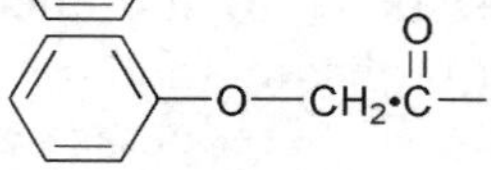

6.28 R = H

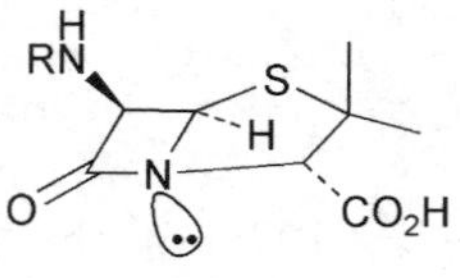

6.29

The chemistry of the penicillins is dominated by the presence of the diminished amide resonance of the β-lactam ring. When this is fused to the thiazolidine ring, the lone pair on the nitrogen cannot participate in amide resonance. Consequently, the β-lactam carbonyl group is sensitive to nucleophilic attack and the nitrogen atom retains some basicity (see **6.29**). Furthermore, the carbonyl group of 6-acyl derivatives can be involved in neighbouring group participation in the cleavage of the β-lactam ring under acid-catalysed conditions (see **6.30–6.31**).

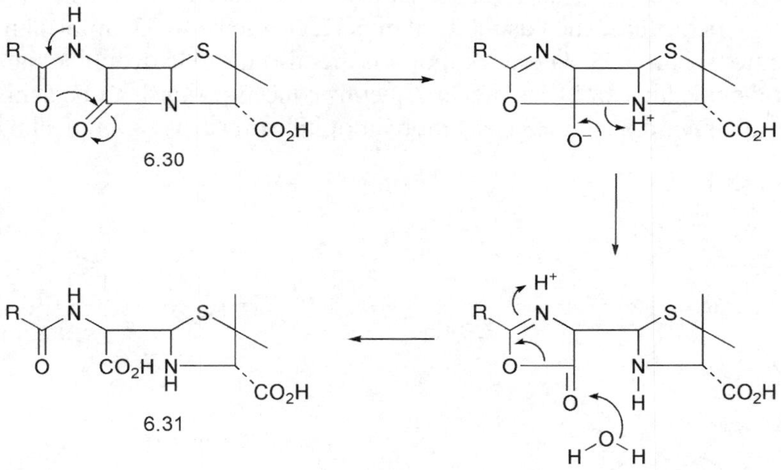

The β-lactam ring, which is essential for biological activity, is also sensitive to enzymatic cleavage by β-lactamases, which are serine proteases. Many amides are hydrolysed in nature by enzyme systems that are serine proteases. The active sites of these enzymes are characterized by the close proximity of the hydroxymethyl group of a serine and the imidazole ring of histidine. The imidazole ring in turn is activated by an adjacent aspartic acid side chain. The imidazole can function as a base abstracting a proton from the serine to create a powerful nucleophile. This can then add to the carbonyl of the amide leading to the cleavage of the amide. The carboxyl unit is attached to the serine from which it is released by a second hydrolytic step also catalysed by a protonated histidine unit. Consequently, the penicillins are readily deactivated. Natural selection has meant that strains of bacteria, which produce β-lactamases survive and hence resistant strains of bacteria have evolved. Bacteria have also evolved efflux resistance mechanisms in which the organism expels the antibiotic.

6.6.1   Semi-Synthetic Penicillins

The availability of 6-aminopenicillanic acid (6-APA) has meant that it was possible to vary the structure of the side chain by coupling different acid chlorides with the 6-amino group. The semi-synthetic penicillins have been designed to reduce the ease with which the cleavage of the β-lactam ring takes place. Groups have been introduced to reduce the effect of neighbouring group participation by making it sterically more difficult for the acyl carbonyl group to add to the β-lactam carbonyl and by introducing amino and carbonyl groups, which alter the basicity and conformation of the side chain. This has led to the development of semi-synthetic penicillins, such as methicillin **6.32**, cloxacillin **6.33**, ampicillin **6.34** and amoxycillin **6.35**. These compounds are also more hydrophilic than the natural penicillins and thus higher aqueous concentrations can be achieved. The ester, pivampicillin **6.36**, is a more lipophilic pro-drug of ampicillin **6.34**.

6.32

6.33

6.34 R′ = R″ = H

6.35 R′ = OH, R″ = H

6.36 R′ = H, R″ = –CH₂–O

6.7   CLAVULANIC ACID AND THE INHIBITION OF β-LACTAMASES

The β-lactamases are inhibited by another microbial product, clavulanic acid **6.37**, which is produced by *Streptomyces clavuligenus* and discovered in 1975. Clavulanic acid can function as an irreversible suicide inhibitor of the β-lactamases because it can bind using two reactive centres to the enzyme. Clavulanic acid **6.37** contains an electron-rich enol–ether. Protonation of this during enzymatic hydrolysis of the β-lactam exposes two electron-deficient carbon atoms, **6.38**, which can add a nucleophile from the enzyme surface thus irreversibly binding clavulanic acid to the β-lactamase. The clavulanic acid has acted as a suicide inhibitor of the β-lactamase. Clavulanic acid, while only a weak antibiotic in its own right, has been co-formulated with amoxycillin **6.35**, to provide a valuable combination known as Augmentin® in which the clavulanic acid inhibits the β-lactamases allowing the penicillin to work.

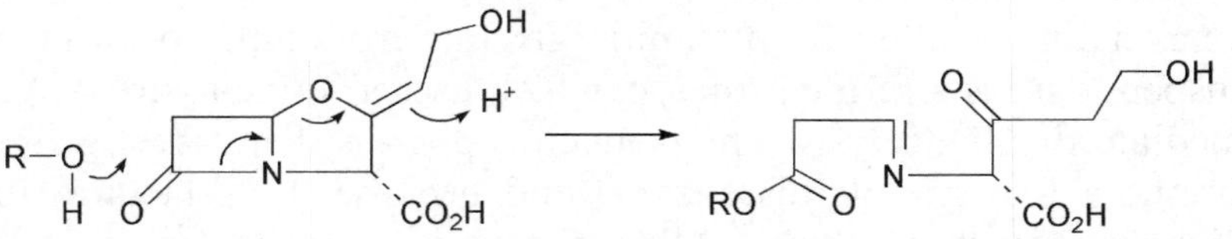

6.8   THE CEPHALOSPORINS

The post-war search for other β-lactam antibiotics led to the discovery of the cephalosporins from another fungus, *Cephalosporium acremonium* (*brotzu*). This fungus was isolated in 1945 from a sewage outfall in Sardinia. Many of the developments made with the penicillins have been extended to the cephalosporins. These include the cleavage of the side chain to form 7-aminocephalosporanic acid **6.39** and the subsequent synthesis of a range of semi-synthetic cephalosporins exemplified by cephalosporidine **6.40**. Many of the active cephalosporins have a leaving group attached to C-3 and this may participate in the cleavage of the β-lactam. Other naturally occurring β-lactam antibiotics include thienamycin **6.41** and norcardicin **6.42**.

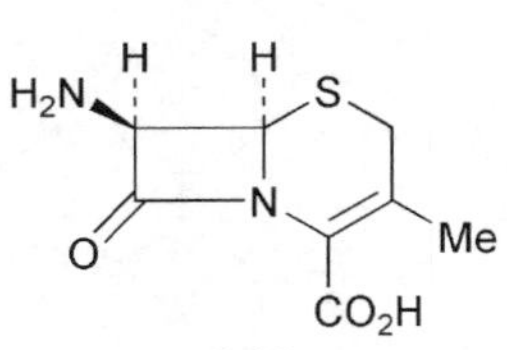

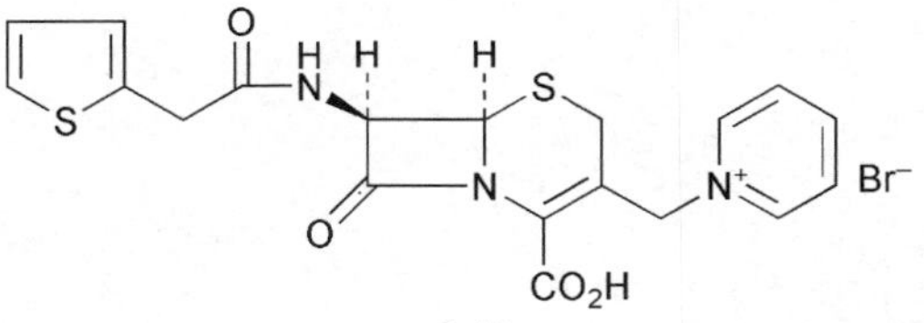

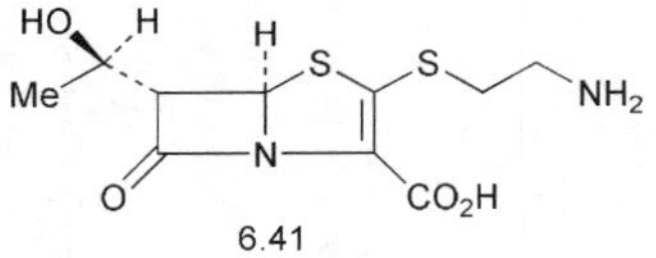

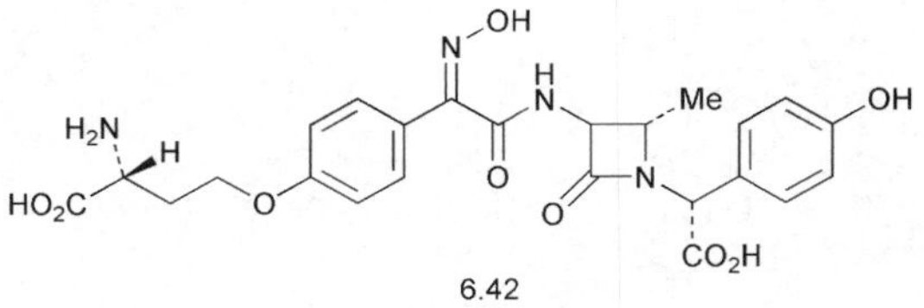

6.9   THE MODE OF ACTION OF THE β-LACTAM ANTIBIOTICS

The biological activity of the penicillins arises from the inhibition of bacterial cell wall biosynthesis. The bacterial cell wall is a peptidoglycan structure in which *N*-acetylmuramic acid **6.43** and *N*-acetylglucosamine

6.44 units are linked together. Attached to these are peptide chains, which are cross-linked to give rigidity. The cross-linking step is mediated by an enzyme, transpeptidase. This cleaves a D-alanine–D-alanine bond in one chain to allow a D-alanine–glycine cross-link to form **6.45**. Transpeptidase is a serine protease, which uses a serine hydroxyl group to mediate this hydrolysis. The penicillins possess the correct geometry to fit the active site of this enzyme and because the β-lactam ring is particularly reactive, they acylate the serine preventing its further catalytic role. Inhibition of the transpeptidase prevents cross-linking and the rigid peptidoglycan structure does not form. Given that bacteria are continually replicating their cell wall, the failure to form this link in the cell wall leads to lysis and cell death. The cephalosporins can also bind to a serine. The 3-acetate can play a role in this as a leaving group.

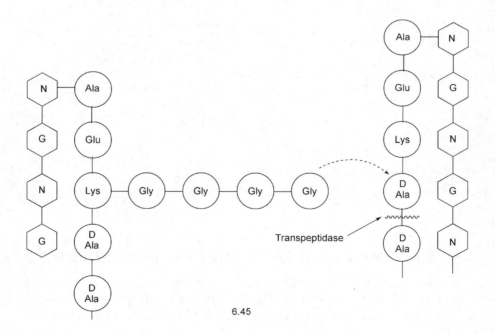

6.45

6.10 OTHER ANTIBIOTICS

Tuberculosis is a serious bacterial disease produced by infection with *Mycobacterium tuberculosis*. In the early 1940s, a research programme initiated by Waksman in the USA investigated the action of soil micro-organisms particularly, *Streptomycetes* against pathogenic micro-organisms. In 1944, an aminoglycoside antibiotic, streptomycin, was isolated from *Streptomyces griseus*. This antibiotic showed significant activity as an anti-tubercular drug and it has made a major contribution to the control of this disease. Its structure was established in 1947.

Some antibiotics produced by *Streptomycetes*, block protein synthesis in bacteria by binding to various units of the ribosome. Thus chloro-tetracycline (aureomycin) **6.46**, isolated from *S. aureofaciens* in 1945, inhibits protein synthesis by binding to the 30S sub-unit of the ribosome. This prevents the aminoacyl tRNA from binding to the ribosome. It was introduced in 1948 and its structure was established in 1952. Chloramphenicol **6.47** and erythromycin **6.48** also affect the bacterial ribosomes. They bind to the 50S subunit and prevent the translocation of peptide groups. Erythromycin **6.48** is an example of a macrolide antibiotic, so called because it possesses a large ring lactone. A complex antibiotic, vancomycin, was isolated from *S. orientalis* in the mid-1950s. Although this antibiotic targets the bacterial cell wall, it does so in a different way to the penicillins. It is a valuable antibiotic because it is active against penicillin resistant strains of *Staphylococci* and it can be used to combat methicillin resistant strains of *Staphylococcus aures* (MRSA).

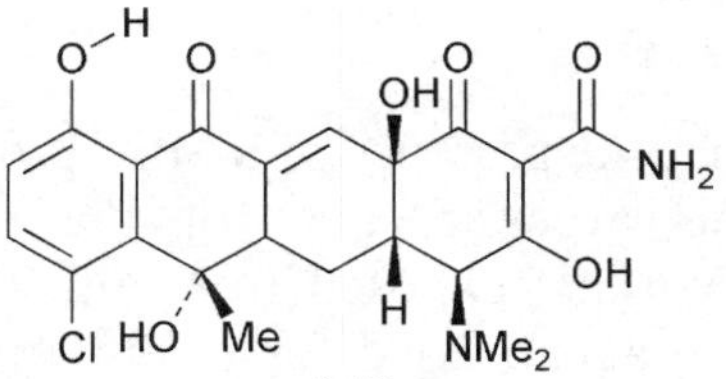

6.46

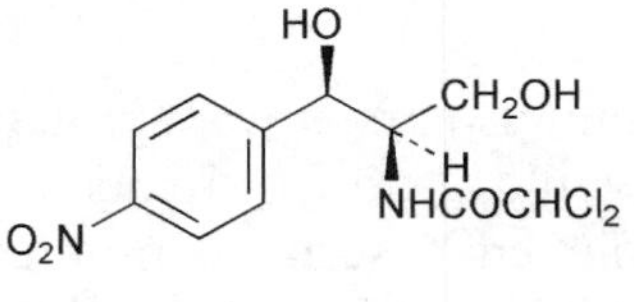

6.47

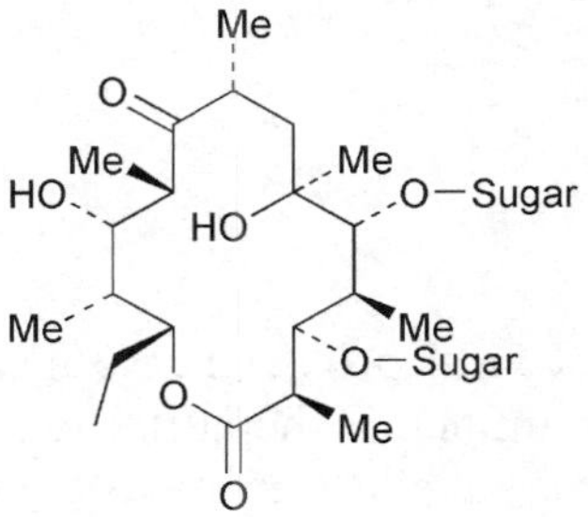

6.48

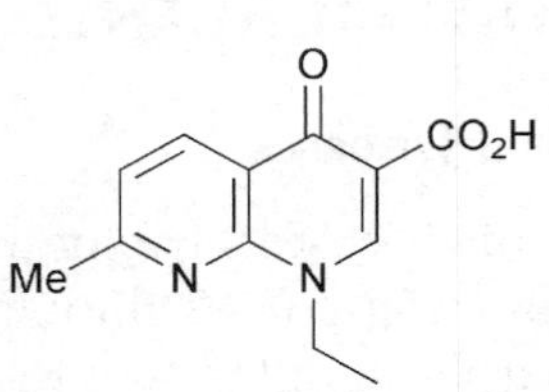

6.49

6.11   SYNTHETIC ANTI-BACTERIAL AGENTS

Some synthetic anti-bacterial agents have been developed which act on nucleic acid transcription and replication. A family of quinolones and fluoroquinolones including nalidixic acid **6.49** and ciprofloxacin **6.50** have been discovered recently, which bind to DNA gyrase and prevent the coiling of DNA to form its tertiary structure. The widespread appearance of MRSA in hospitals has provided the stimulus for the synthesis of novel anti-bacterial compounds. A new class of anti-bacterial agents have been developed that contains an oxazolidinone ring and which inhibit protein synthesis in bacteria. These include linezolid **6.51** which was introduced as Zyvox® in 2001.

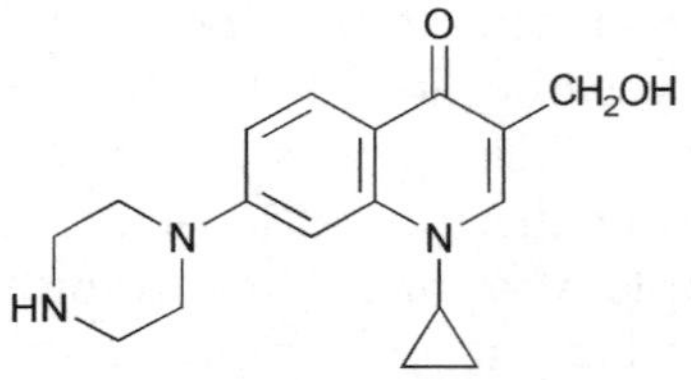

6.50 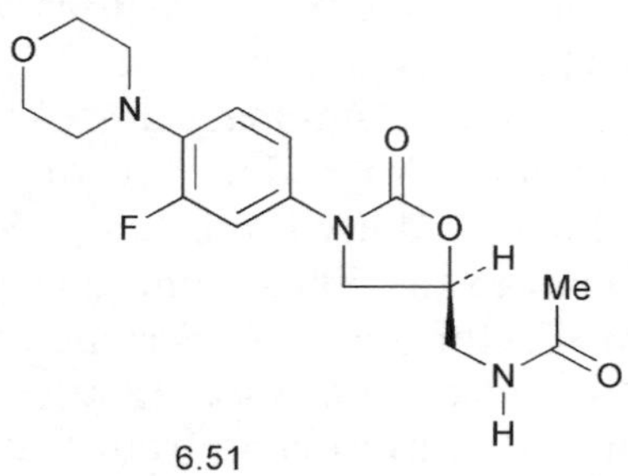   6.51

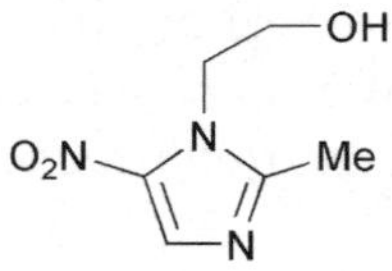

6.52

For many years, it was thought that the presence of a nitro group in a drug would lead to toxicity. This arose because of the methaemoglobinaemia seen in munition workers exposed to trinitrotoluene. However, a number of anti-bacterial agents used for the treatment of urogenital infections such as nitrofurantoin and metronidazole **6.52** have been developed which contain this group.

6.12   ANTI-VIRAL AGENTS

6.12.1   Viral Diseases

A number of diseases of man are of viral origin. These include poliomyelitis, common cold (rhinovirus), influenza, hepatitis A and B (liver disease), herpes simplex (cold sores), rubella and measles, papillomas (warts) and the human immunodeficiency virus (HIV-AIDS). A virus can rapidly mutate and hence different strains of the same virus can appear. These can have a different pathogenicity and immunity acquired against one strain may not be effective against another. This is common experience with the influenza virus.

6.12.2 Viral Structure and Replication

The structure of the viral particle (viron) provides the basis for the treatment of viral infections. The virus **6.53** comprises a central core of nucleic acid material (either DNA or RNA), protein containing the polymerase enzyme involved in viral replication, a protein coat and the surface glycoprotein. The latter, comprising a haemagglutinin and a neuraminidase, are involved in the recognition and interaction between the virus and its host cell. In the various strains of the 'flu virus', there are up to 15 different types of haemagglutinin and nine different neuraminidases. Thus, the Asian flu strain of 1957 was the H2N2 strain, while the avian flu strain which is currently of concern is H5N1.

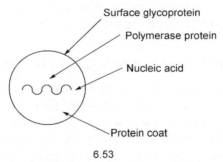

6.53

In the course of a viral infection, the virus particle becomes attached to the host cell-surface via a receptor-surface protein interaction. A vacuole is formed by the host cell around the virus particle and the virus is transported into the cell where it releases its core DNA or RNA material. These replicate within the cell using the host cell's replication machinery. An assembly process takes place and new viral particles are formed and expelled from the host cell to infect other cells. In the replication of a DNA virus, transcription of the viral DNA to form mRNA occurs leading to the synthesis of viral protein. With an RNA virus, the enzyme system reverse transcriptase is required to synthesize viral DNA from the viral RNA template. This viral DNA may then replicate using the host-cell enzymes. Transcription of the new viral DNA regenerates the viral RNA and the viral protein for the new progeny.

The fact that a virus requires a host cell in which to replicate makes the bioassay of anti-viral compounds more difficult than that of anti-bacterial agents. Furthermore, since a virus can mutate rapidly, a treatment which is effective against one strain of, for example the 'flu virus,' may not be effective against another strain.

The immune system in the body may recognize the surface glycoproteins on the virus. Antibodies are formed which encapsulate and expel the virus. Vaccination uses an attenuated or weakened form of the virus, which nevertheless stimulates the formation of the relevant antibodies. The immune system is then prepared to cope with an invasive attack from the 'wild-type' virus. The consequences of the immune reaction include the fever and rhinitis associated with viral infections.

6.12.3   Targets for Anti-Viral Agents

The viral cell cycles reveal several potential targets for anti-viral chemotherapy. Nucleic acid synthesis is central to viral replication and therefore forms a key target. The enzyme systems reverse transcriptase and DNA polymerase also play important roles in the development of the virus and are suitable targets. The surface glycoprotein and its synthesis by the virus represent a further target.

6.13   THE INHIBITION OF NUCLEIC ACID BIOSYNTHESIS

Nucleic acid biosynthesis involves the coupling of the pyrimidine and purine bases with ribose or deoxyribose to form the nucleoside, phosphorylation of the sugar to form the nucleotide and polymerization to give the nucleic acid. Nucleotides and their analogues are too polar to cross the cell membrane to reach the site of viral replication. However, a number of nucleoside analogues are useful anti-viral agents. These rely on the viral kinase to phosphorylate them once they have crossed the cellular membrane and reached the site of viral replication. Because rapid nucleic acid biosynthesis is also a characteristic of cancer cells, a number of nucleoside analogues with anti-viral properties have also attracted interest as tumour inhibitors.

Acyclovir (Zovirax®) **6.54** is a guanosine analogue which lacks part of the deoxyribose. Once it has entered the virus, the viral kinase phosphorylates acyclovir to give the triphosphate and attaches this to the developing nucleic acid chain. However, further chain elongation cannot take place because the relevant sugar hydroxyl group is missing. This anti-viral drug is relatively safe and it is used in the treatment of herpes simplex virus. Acyclovir is found in some 'over-the-counter' medicines for the treatment of 'cold sores'. A pro-drug, valaciclovir **6.55**, has a valine unit attached, while a carbocyclic analogue, carbovir **6.56**, is also used.

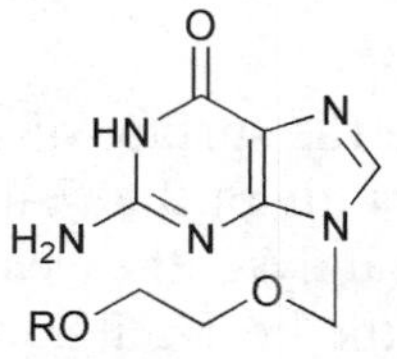

6.54 R = H

6.55 R = —C—CHCH
              ‖ | \
              O NH₂ Me

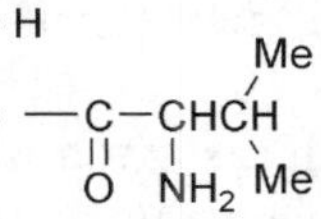

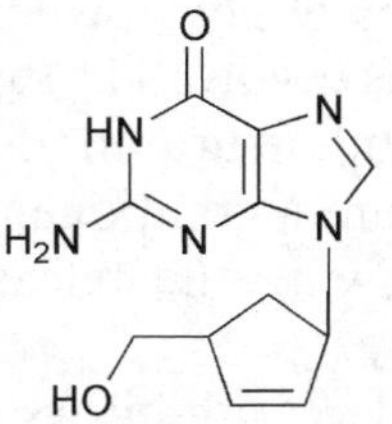

6.56

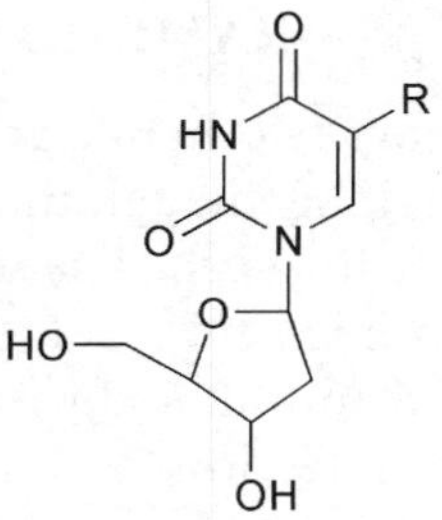

6.57 R = I
6.58 R = CF₃

The thymidine analogues, idoxuridine **6.57** and trifluorothymidine **6.58**, are phosphorylated to produce inhibitors of the viral DNA polymerase. However, they are also phosphorylated by the host kinases in uninfected cells and hence they are less selective and consequently they are less widely used as anti-viral agents.

6.14   INHIBITORS OF REVERSE TRANSCRIPTASE

Another nucleoside analogue is azidothymidine (AZT) **6.59**. This compound blocks reverse transcriptase in a RNA virus. AZT is particularly useful in the treatment of HIV-AIDS in which it prevents the HIV from expressing its genome. The human immunodefiency virus is the causative agent for AIDS. This autoimmune deficiency is associated with a depletion of the CD4⁺ lymphocytes which allows various opportunistic infections such as pneumonia to eventually develop and kill the patient. As the HIV infects the CD4⁺ cells, the reverse-transcribed RNA genome of HIV becomes permanently integrated into the host-cell genome. Consequently, reverse transcriptase is a major target for anti-AIDS drugs. There are a number of other reverse transcriptase inhibitors including neviripine **6.60**. Some adamantanes such as amantadine **6.61**, interfere with the process of uncoating viral particles in the cell to release their own DNA or RNA.

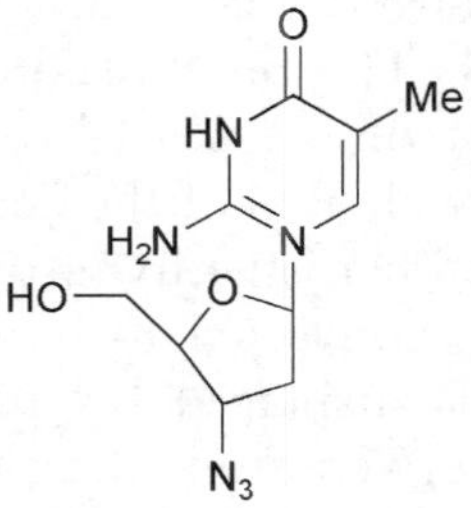

6.59

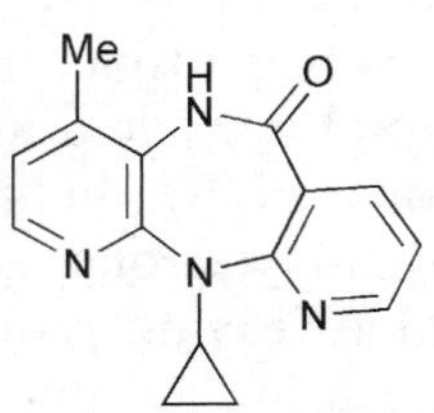

6.60

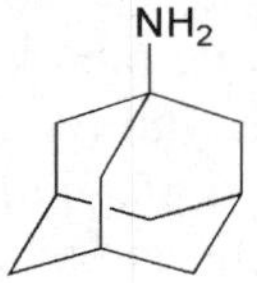

6.61

6.15   NEURAMINIDASE INHIBITORS

There are two major groups of glycoprotein on the surface of the virus. The first, haemagglutinin, is involved in the docking of the virus particle with the host cell. The second, neuraminidase, catalyses the cleavage of a neuraminic acid residue from a component of the host-cell gangliosides in its membrane thus allowing the release of new virus particles for further infection.

A group of neuraminidase inhibitors such as zanamivir **6.62** are attracting interest because they inhibit the formation of this glycoprotein in the viral cell-wall covering. The particular area of application is in the treatment of influenza. Zanamavir (Relenza®) **6.62** is active against both influenza A and B. It is modelled on neuraminic acid and binds to the active site of neuraminidase with significant hydrogen-bonding interactions to arginine and glutamic acid residues on the enzyme surface. Oseltamivir (Tamiflu®) **6.63** is another compound of this type, which is attracting interest because it is active against avian flu. It is synthesized from the natural product, shikimic acid.

6.62   6.63

6.16   THE SYNTHESIS OF NUCLEOSIDE ANALOGUES

The syntheses of these nucleoside analogues reveal a number of aspects of the chemistry of pyrimidines and purines. The pyrimidine ring system of thymidine is synthesized by the condensation of urea and methylmalondialdehyde. The N–H of the pyrimidine is rendered weakly acidic by the carbonyl groups and hence the amide reacts with a protected chlorodeoxyribose to form thymidine **6.64**. The preparation of AZT **6.59** from thymidine involves displacement of the 3′-hydroxyl group with retention of configuration. The primary alcohol of the deoxyribose of thymidine **6.64** is protected by a bulky triphenylmethyl group allowing the secondary 3′-hydroxyl group to be converted to its methanesulfonate **6.65**. This good leaving group is displaced by an internal nucleophilic substitution using an anion derived from the pyrimidine ring. However, the pyrimidine ring of the anhydrothymidine **6.66** can also behave as a leaving group and is displaced by the azide anion to give

AZT 6.59. This double inversion leads to the azido group taking up the same configuration as the 3′-hydroxyl group of thymidine.

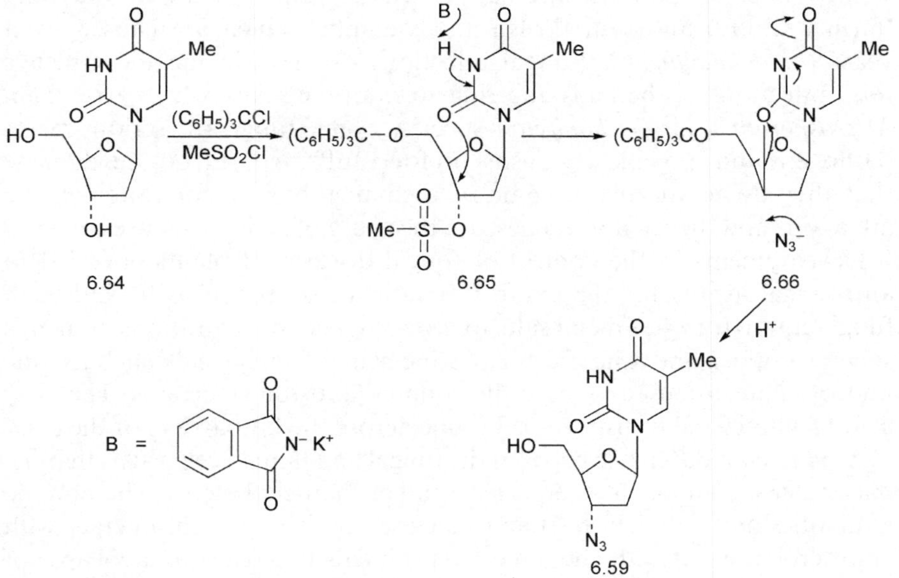

6.64   6.65   6.66

B =

6.59

The 9-N–H of guanine **6.67** is rendered acidic by the imino group of the purine ring system. In some respect, it is the nitrogen analogue of a carboxylic acid. A protected guanine treated with triethylamine and the chloromethyl ether **6.68** gives the benzoate of acyclovir **6.69**, which is then hydrolysed to acyclovir **6.54**.

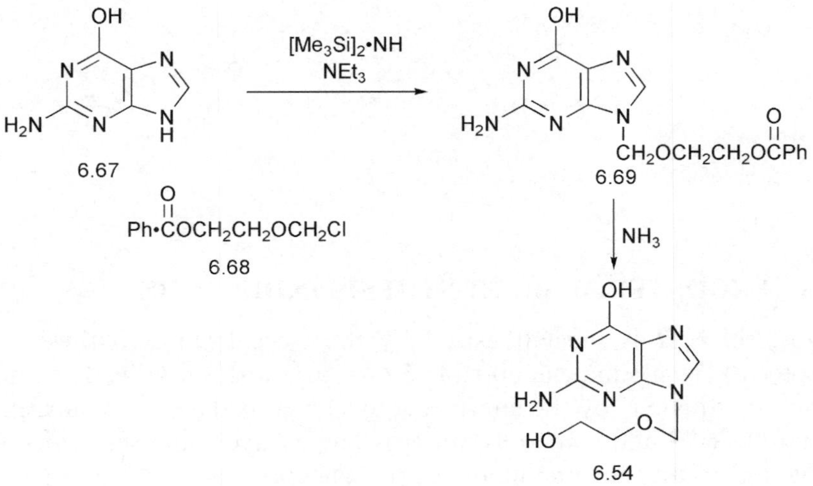

6.67

$Ph\cdot COCH_2CH_2OCH_2Cl$

6.68

6.69

6.54

6.17 ANTI-FUNGAL AGENTS

Fungal and yeast infections in man occur mostly on the skin. A few serious infections occur internally such as Aspergillosis in the lung (farmer's lung) and oral thrush and vaginitis, which are caused by a yeast *Candida albicans*. Fungal infections of the skin include athlete's foot, infections of the nails and ringworm in the scalp, which arise from *Microsporum* and *Trichophyton* species. Some fungi are serious plant pathogens and spoilage organisms on foodstuffs. Apart from the damage that they do to the plant, some of their metabolites, for example, the aflatoxins and the trichothecenes, are the cause of serious diseases in man.

Developments in the control of fungal diseases of plants have led to anti-fungal agents for the control of fungal diseases in man. Although fungi require traces of metal salts to grow, higher concentrations of metals such as copper and zinc are toxic. Zinc salts of fatty acids such as zinc undecylenate, are used in 'over-the-counter' anti-fungal creams. The fatty acid facilitates the transport of the zinc across the fat barriers of the skin.

A significant difference between the fungal and human cell wall is that the major sterol component is ergosterol rather than cholesterol. The polyene antibiotics amphotericin **6.70** and its close relative nystatin, interact with ergosterol and not with cholesterol. This interaction leads to a collapse of the cell-wall structure and to a leakage of the fungal cell wall constituents.

6.70

6.18 ERGOSTEROL BIOSYNTHESIS INHIBITORS

Ergosterol **6.72** is biosynthesized by the typical sterol pathway (see chapter 5) from squalene epoxide **5.64** via lanosterol **6.71**. The epoxidation of squalene by the enzyme, squalene epoxidase, is the target for tolnaftate **6.73** and some allylamines. These have some selectivity and show less affinity for mammalian squalene epoxidase.

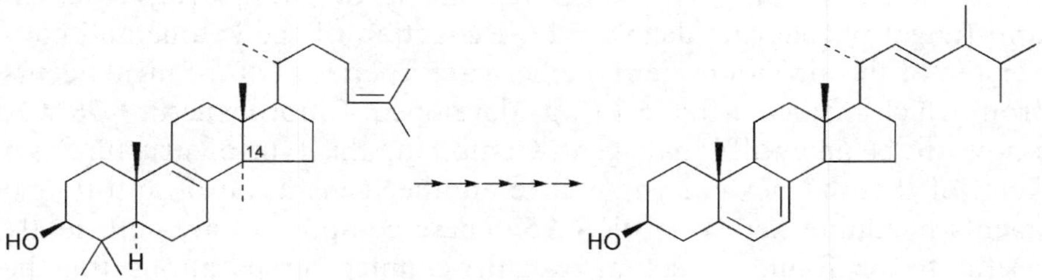

6.71

6.72

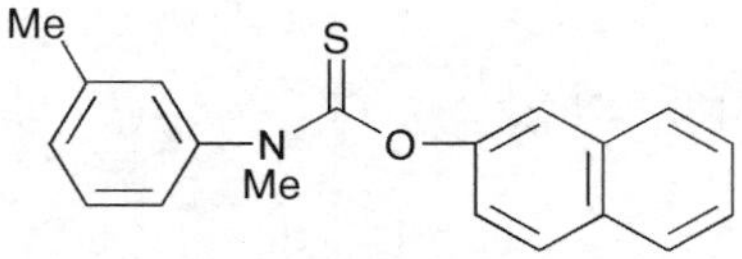

6.73

An important step in the biosynthesis of ergosterol involves the removal of the 14α-methyl group from lanosterol. The enzyme system that mediates this is a cytochrome P_{450} dependent system. The 14α-methyl group is hydroxylated and oxidized to an aldehyde before being removed as formic acid. The azole fungicides, such as micoconazole **6.74**, ketoconazole **6.75**, and fluconazole **6.76**, contain an imidazole ring or in more recent examples, a triazole ring.

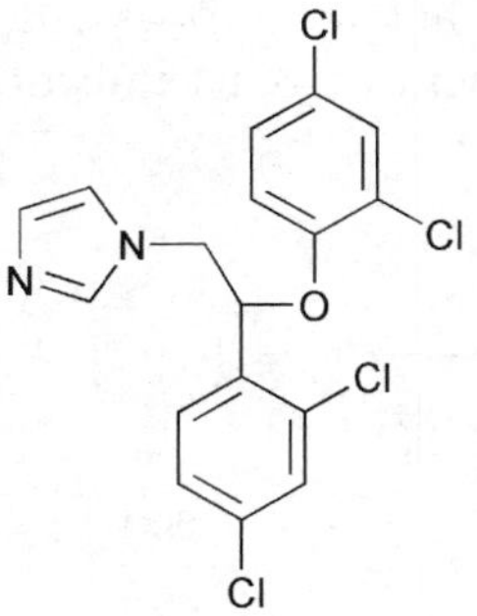

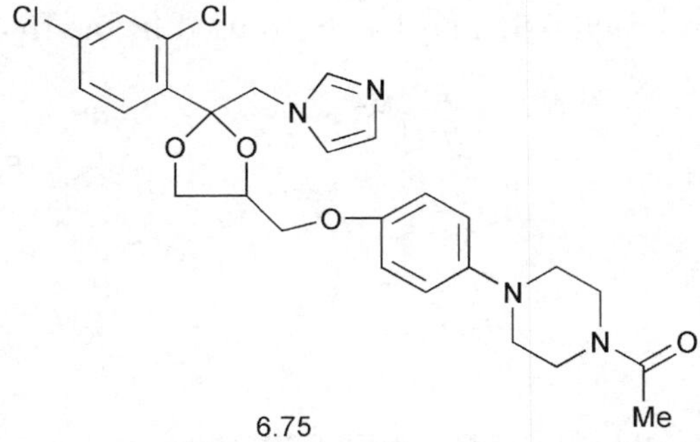

6.74

6.75

This co-ordinates to the iron in the cytochrome preventing it from delivering the oxygen. Since cytochrome P_{450} enzyme systems are involved in many other biosynthetic oxidations, the remainder of these rather complex molecules is present to ensure selectivity for this particular enzyme system. Other azoles are used, for example, as plant growth regulators in which they inhibit the cytochrome P_{450} mediated oxidation of ent-kaurene, a precursor of the gibberellin plant-growth hormones. Azoles are also used in the treatment of breast cancer (see Chapter 7).

The starting point for the development of these drugs was an anti-fungal phenacylimidazole **6.77**. Reduction of the ketone and conversion of the alcohol to benzyl ether gave a series of anti-fungal agents from which micoconazole **6.74** was developed. Chlorphenesin **6.78** was known to be an anti-fungal agent. Combining this type of structure as a ketal of the phenacylimidazole gave another generation of anti-fungal agents including ketoconazole **6.75**. These compounds are sufficiently useful to be found in some 'over-the-counter' preparations for the treatment of athlete's foot and fungal infections of the scalp.

6.76   6.77   6.78

6.19 OTHER ANTI-FUNGAL AGENTS

The anti-fungal antibiotic, griseofulvin **6.79** has been used in the treatment of a number of fungal infections, particularly of the nails and of the scalp. Its action may be to interfere with micro-tubule function in the fungus. Some of the nucleoside antibiotics, which interfere with DNA synthesis have anti-fungal activity. Flucytosine **6.80** can act as a pro-drug for fluorouracil **6.81** and has been used in the treatment of yeast infections.

6.79   6.80   6.81

6.20 PARASITIC INFECTIONS

6.20.1 The Treatment of Malaria

Malaria is a serious protozoal infection of man. It is caused by four species of protozoa, *Plasmodium falciparun*, *P. vivax*, *P. malariae* and *P. ovale*. These protozoa also spend part of their life cycle in the female *Anopheles* mosquito. When the mosquito bites a human, it injects sporozoites of the protozoa into the blood stream. These are carried to the liver where they multiply to form tissue schizonts. After some days these schizonts rupture and release merozoites which infect red blood cells. This erythrocyte stage of the disease produces the characteristic fever. When a mosquito then bites an infected person, it sucks up blood containing the parasite and so spreads the disease.

Most anti-malarial drugs such as the alkaloid quinine **6.82** from Cinchona bark, and the synthetic compound chloroquine **6.83** act on the erythrocyte schizants. Pyrimethamine **6.84** and proguanil **6.85** are slower-acting drugs, which are folate antagonists that also act on the parasites. Proguanil is metabolized to a cycloguanil derivative **6.86**, which is the active drug. This metabolite has a marked similarity to pyrimethamine **6.84**. Both compounds are dihydrofolate reductase inhibitors. Proguanil is also given in combination with a quinone, atavaquone **6.87**. A combination with dapsone (Lapdap®) is also used.

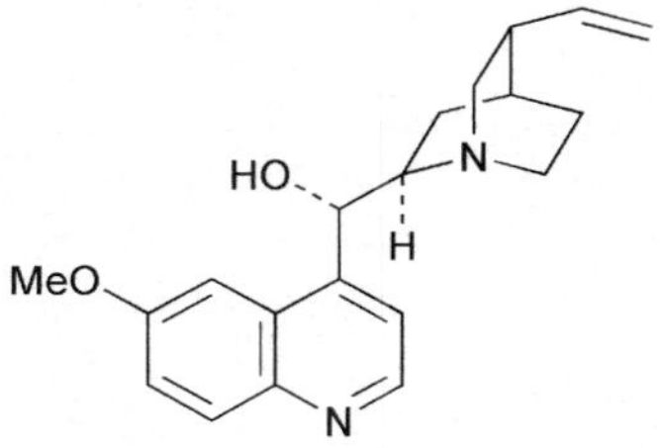

6.82

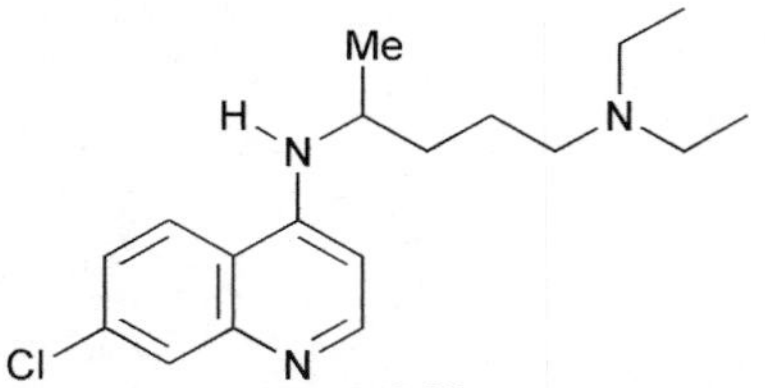

6.83

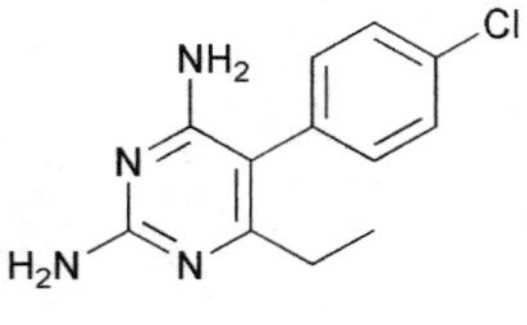

6.84

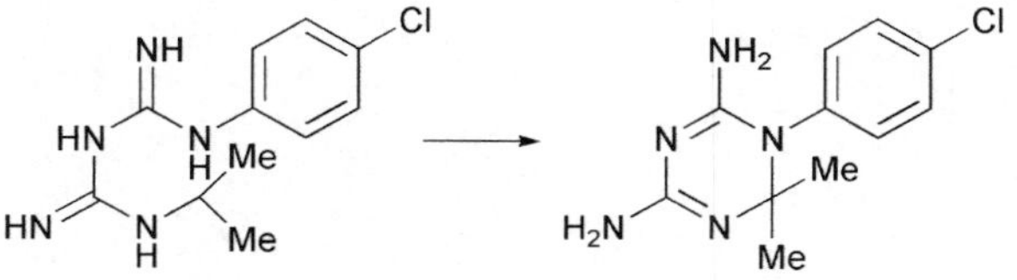

6.85

6.86

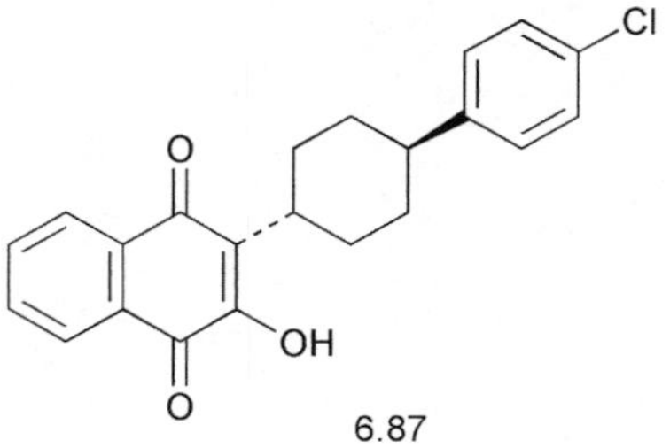

6.87

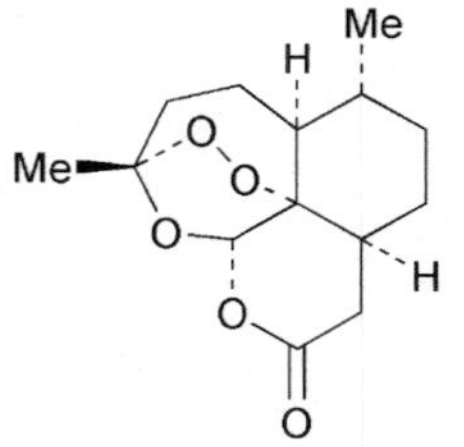

6.88

Some resistant forms of the malaria parasite have developed and the search for new drugs has been undertaken. An unusual sesquiterpenoid, artemisinin **6.88** from the Chinese plant *Artemisia annua* (Qinghaosu) has provided useful leads. The spiro-ether, artemether **6.89** has been developed. The alkaloid, febrifugine **6.90** from the Chinese drug changs-hen, has formed the basis of some further studies in this area.

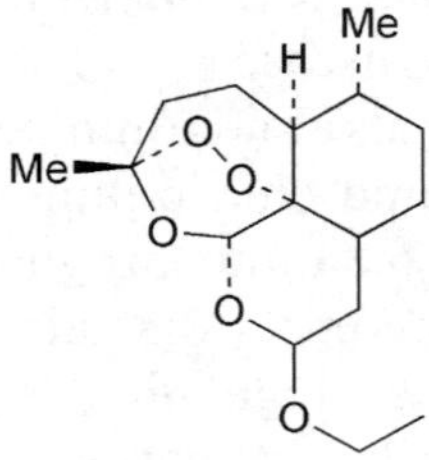

6.89 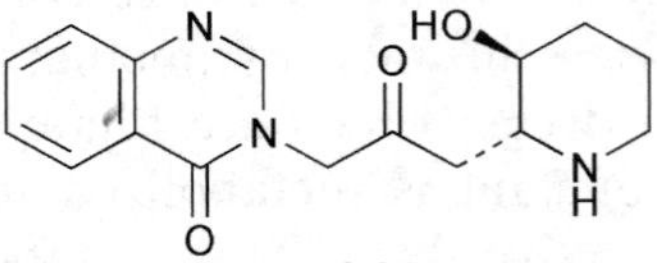   6.90

CHAPTER 7

Cancer Chemotherapy

7.1   AIMS

The aim of this chapter is to show how cancer chemotherapy exploits the subtle differences between the normal and the cancerous cell. By the end of the chapter you should be aware of:

- the stages in the development of the cell;
- the role of anti-metabolites in blocking nucleic acid biosynthesis;
- the role of alkylating and intercalating agents in interfering with nucleic acid function;
- the role of anti-mitotic agents in interfering with cell division; and
- the role of specific inhibitors of particular developmental processes such as aromatase inhibitors of estrogen biosynthesis in the treatment of breast cancer.

7.2   INTRODUCTION

Cancer is one of the major causes of death and one of the most feared of diseases. The term covers a range of diseases of multi-cellular organisms linked by the common feature of an abnormal, poorly regulated and often invasive growth of the organism's own cells. The tumourous growth of tissue may be benign or it may be malignant and producing secondary growths known as metastases. Chemotherapy is one of a number of treatments for cancer that also include surgery and radiation and it is often used in combination with these.

7.3   THE CELL CYCLE

The cycle of development of a cell involves four main phases (see Scheme 1). The duplication of the DNA takes place in the synthesis

129

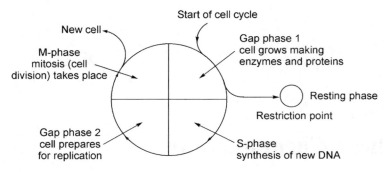

Scheme 1   *The cell cycle*

(S) phase and the separation into two cells (mitosis) occurs in the M phase. There are two other phases known as gaps, G_1 preceding the S phase and G_2 preceding the M phase during which other cellular structures are synthesized. Normal cells may enter a resting phase (G_o) from the G_1 phase. However, cancerous cells continue to divide and do not necessarily enter a resting state at this stage in the cycle. The progress of a cell through these stages is mediated by proteins known as cyclins. These activate cyclin-dependent kinases (CDKs), which set the cell signalling pathways in motion. There are two sets of genes that control the progression. They are the proto-oncogenes, which control the growth of cells and the tumour suppressor genes, which control the DNA transcription.

A cancerous cell differs from a normal cell in that the factors that regulate its growth and prevent it from invading adjacent cells, are not operative. Whereas normal cells will anchor themselves to other cells, cancerous cells do not need to anchor themselves and continue to grow in an unrestrained manner. Adhesion molecules prevent the excessive growth of cells under normal conditions. Apoptosis or programmed cell death does not take place in cancerous cells. A malignant tumour gives rise to secondary growths elsewhere in the body, for example, in the lymphatic system.

There are a number of changes to the nucleic acid material that distinguish a cancerous cell from a normal cell. Some of these involve the incorporation of new genetic or viral material into the DNA. The changes to the nucleic acid material may be brought about by chemical means, which also include interactions with carcinogens such as β-naphthylamine, benzopyrene derivatives and natural products such as the aflatoxins.

Useful drugs for cancer chemotherapy have to exploit the small differences between cancerous and normal cells. Many drugs target various aspects of DNA synthesis and hence act during the S phase of the cell cycle. Others interfere with key enzymes in the development of the cell, or with their expression by the DNA of the cell or with mitosis. Since these targets are common to normal cells, there is often a fine balance between the efficacy and the toxicity of a drug. Side effects in cancer chemotherapy are quite common.

7.4   CANCER CHEMOTHERAPY

It is possible to consider substances used in cancer chemotherapy under a number of headings. These are:

(i) Anti-metabolites: These are compounds that block essential cellular processes leading to the formation of DNA.
(ii) Alkylating agents: These are compounds that react with the nucleic acid bases and prevent them from expressing their code.
(iii) Intercalating agents: These are compounds that become inserted between the strands of DNA leading to breaks in the DNA and preventing replication.
(iv) Polymerase inhibitors: These are compounds that interfere with the enzyme systems involved in the construction and repair of DNA.
(v) Cytotoxic antibiotics: These antibiotics that target rapidly dividing cells, can provide useful tumour inhibitory substances.
(vi) Compounds that affect the development of particular cells, for example, in the breast or prostate gland, provide compounds for the chemotherapy of specific tumours.

7.5   ANTI-METABOLITES

DNA biosynthesis involves the formation of the nucleic acid bases followed by the attachment of the sugar and phosphate units, leading to the formation of the nucleoside and nucleotide, respectively. This is followed by a polymerase reaction in which the nucleic acid chain is formed. Various enzyme systems such as the topoisomerases are then involved in the coiling and folding.

A key step in the formation of the nucleic acid base, thymine **7.2**, is the methylation of the pyrimidine, uracil **7.1**. This methylation forms the target for a number of drugs. Tetrahydrofolic acid is a co-enzyme for the transfer of the C_1 unit in this methylation by the enzyme thymidilate synthetase. Tetrahydrofolic acid is formed by reduction of dihydrofolic acid. Inhibition of the enzyme system, dihydrofolate reductase, by the folic acid analogue, methotrexate **7.4**, prevents the formation of tetrahydrofolic acid and therefore blocks the methylation of uracil. The amino group of the pteridine ring of methotrexate **7.4** facilitates the binding of the drug to the dihydrofolate reductase while the N-methyl group prevents the alkylation reaction that the co-enzyme carries out.

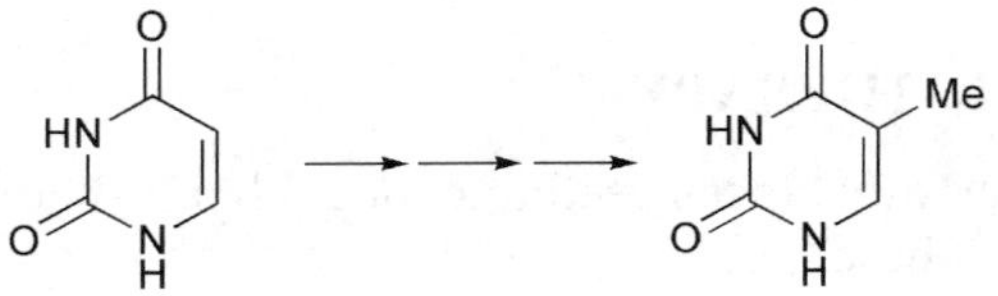

5-Fluorouracil **7.3** possesses a fluorine atom in place of the hydrogen atom of uracil **7.1**, which is replaced by a methyl group in thymidilate synthetase. 5-Fluorouracil is converted to its deoxynucleotide, which binds to thymidilate synthetase where it competes with deoxyuridine monophosphate and blocks the action of the enzyme. The transfer of a C_1 unit from the N^{10}-methylenetetrahydrofolic acid to uracil to form thymine involves the following steps (see Scheme 2). The dissociation of the methylenated uracil from the co-enzyme involves the loss of a proton. In 5-fluorouracil this proton has been replaced by a fluorine atom and the dissociation cannot take place. Hence the 5-fluorouracil proceeds part way along the pathway and then acts as a suicide inhibitor. Capecitabine **7.5** is a fluoropyrimidine pro-drug that is converted to 5-fluorouracil by a series of enzymatic steps, the last of which, hydrolysis by thymidine phosphorylase, takes place predominantly in rapidly dividing tumour cells. It thus has some selectivity.

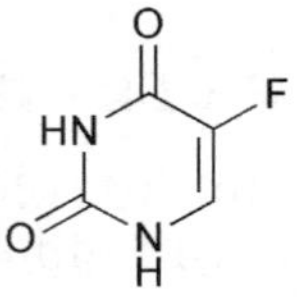

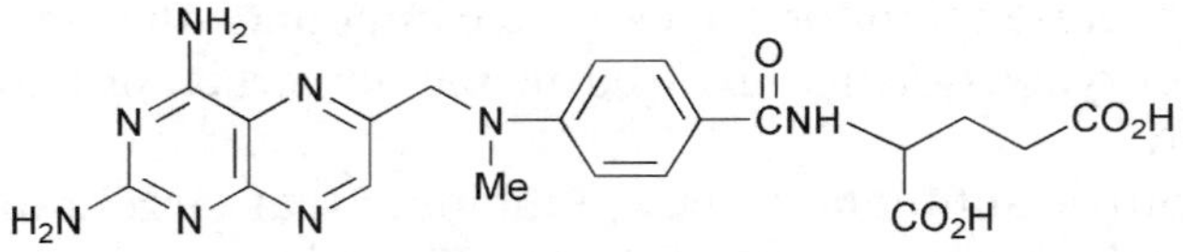

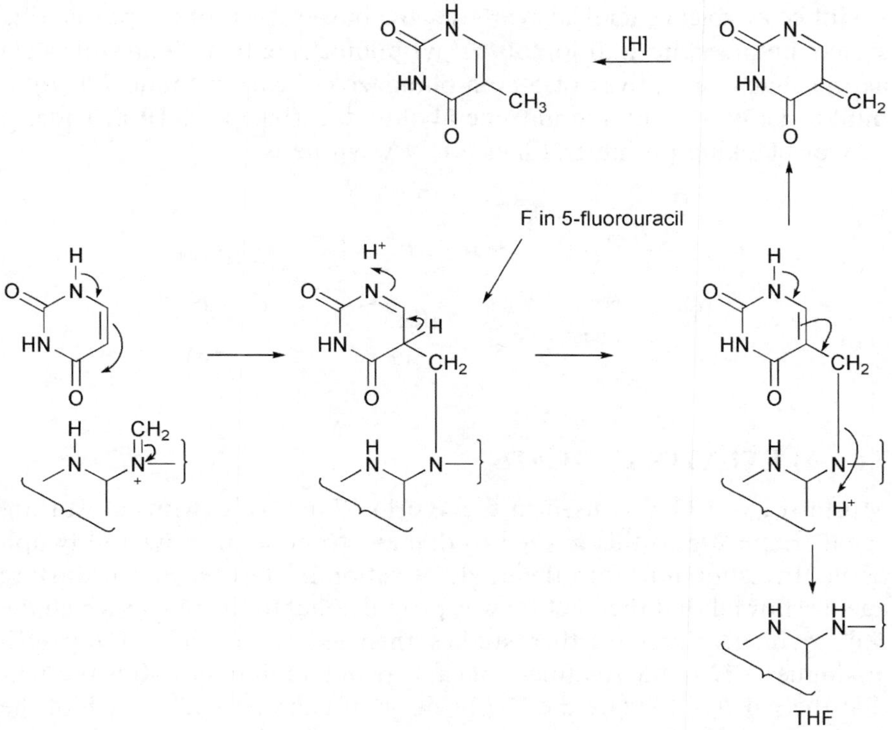

Scheme 2 *The methylation of uracil*

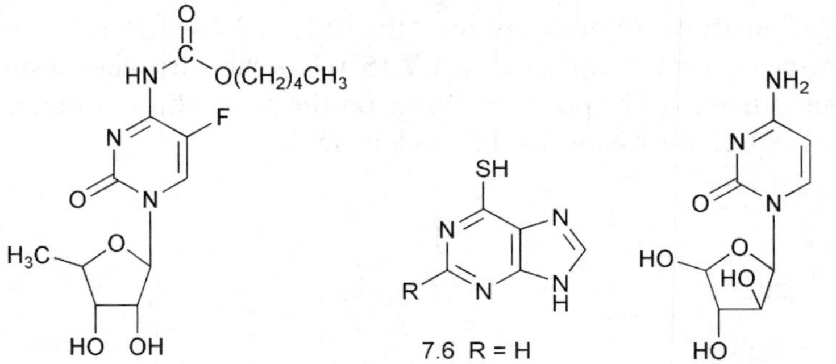

7.5

7.6 R = H
7.7 R = NH$_2$

7.8

6-Mercaptopurine **7.6** and 6-thioguanine **7.7** are analogues of purine bases. They are converted to the corresponding nucleotides such as 6-thioinosinic acid, which interfere with DNA and RNA syntheses and function. Cytarabine (cytosine arabinoside) **7.8** is an analogue of 2'-deoxycytidine in which the ribose sugar is replaced by arabinose. Arabinose has a different stereochemistry at C-2. Incorporation into DNA blocks DNA polymerase and terminates chain elongation.

Unlike a typical chemical synthesis, the biosynthesis of the purine ring system involves the addition of the pyrimidine ring to an imidazole such as the sugar derivatives of 5-aminoimidazole-4-carboxamide **7.9**. Inhibition of this step by the anti-metabolite dacarbazone **7.10** provides a way of blocking purine and hence DNA synthesis.

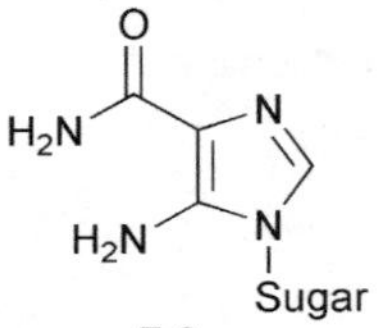

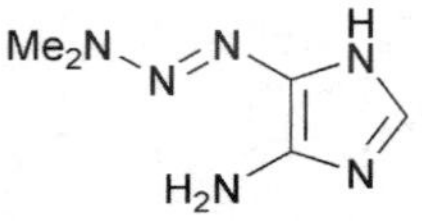

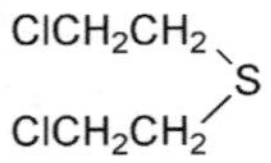

7.6   ALKYLATING AGENTS

Mustard gas **7.11** was used in the World War I as a chemical warfare agent. Exposure to this gas led to damage to bone marrow and lymph tissue. Investigations into the mode of action of these chemical warfare agents showed that their action was proportional to the rate at which the cells were dividing. Further studies then led to the use of nitrogen analogues **7.12** in the treatment of cancers of lymph tissues. It is thought that these drugs alkylate the C-7 position of guanine **7.13** in each of the double strands of DNA bringing about cross-linking and thus interfering with the separation of the strands during mitosis. The reactivity of the chlorine atoms to nucleophilic substitution **7.14–7.16** is enhanced by neighbouring group participation **7.15** from the nitrogen atom of the nitrogen mustard. The positive charge on the N^7 of the guanine may also bring about chain fission in the nucleic acid.

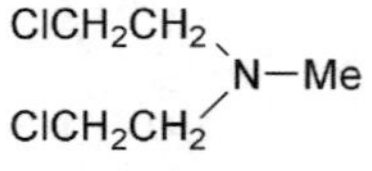

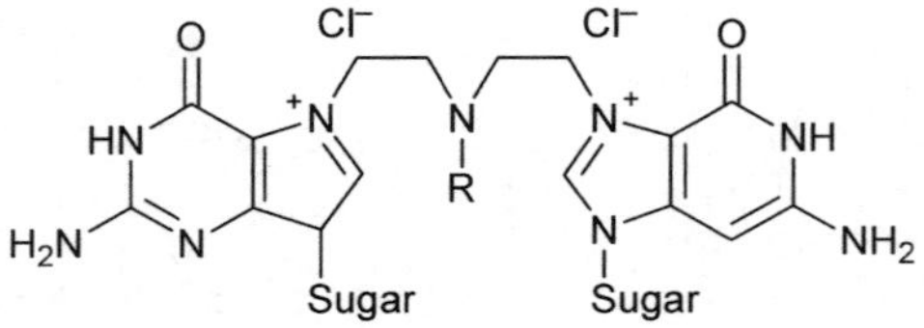

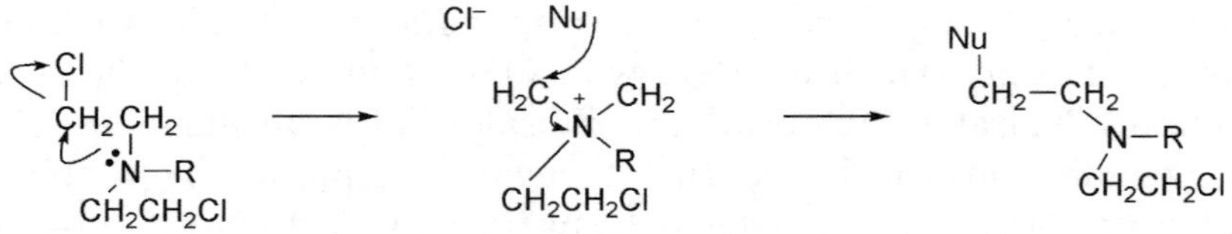

The earliest nitrogen mustard was the simple bis-(2-chloroethyl)amino compound **7.12** but this has been replaced by more selective compounds such as chlorambucil **7.17** and the phenylalanine mustard, melphalen **7.18**

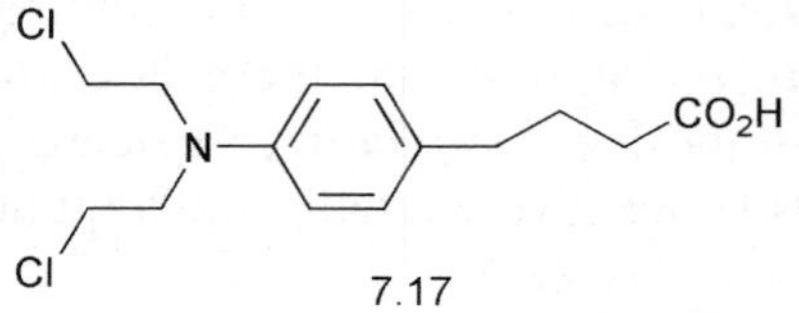

7.17

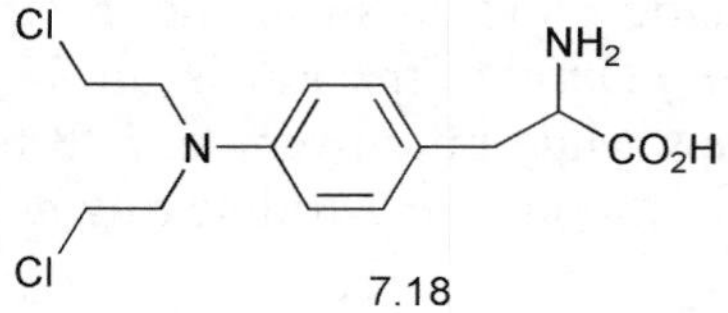

7.18

Cyclophosphamide **7.19** is an interesting pro-drug for a nitrogen mustard. Oxidation of the ring gives 4-hydroxycyclophosphamide, which then fragments via **7.21** to give the active phosphoramide **7.22**.

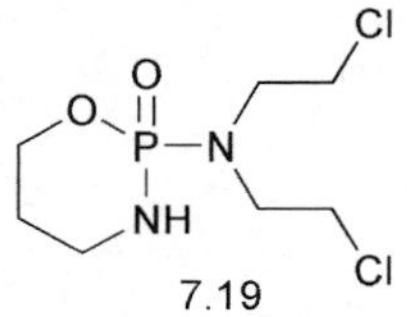

7.19

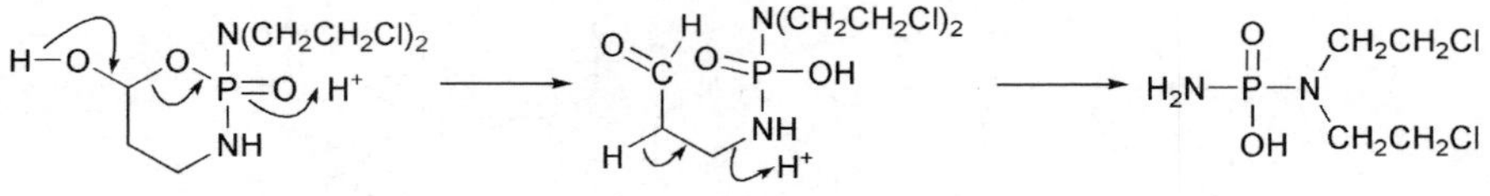

7.20

7.21

7.22

The aziridine ring that is involved as an intermediate **7.15** in the neighbouring group participation by the nitrogen atom in the displacement of the chlorine atom in the nitrogen mustard, is embodied in the alkylating agent, thiotepa, triethylenethiophosphoramide **7.23**. Nitrogen mustards have been attached to steroids to target specific organs such as the prostate gland. An example is estramustine **7.24** in which the carrier steroid is ethynylestradiol. This behaves as a pro-drug releasing the nitrogen mustard in the target organ.

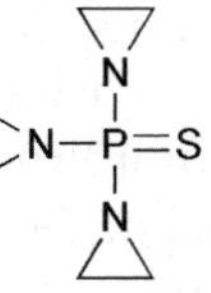

7.23

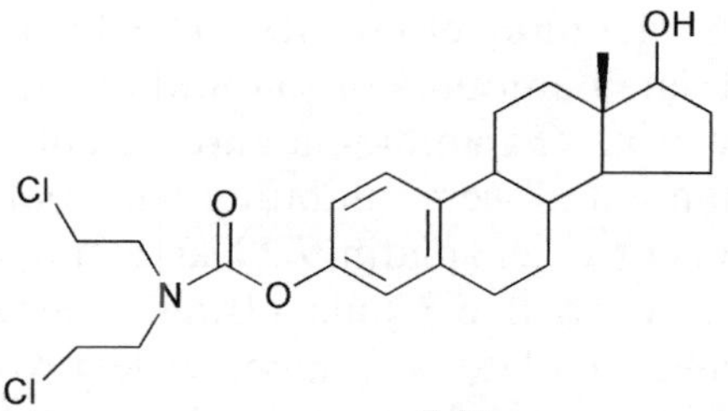

7.24

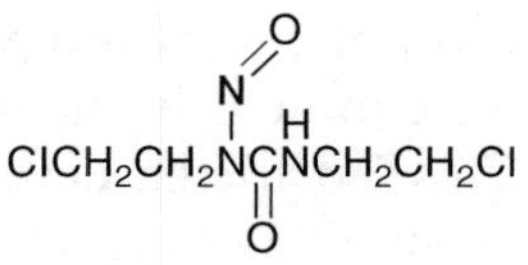

7.25

Another family of reactive alkylating agents is the chloroethylnitroso-ureas exemplified by carnustine **7.25** and lomustine **7.26**. These nitroso-ureas can fragment to yield two reactive fragments both of which can attach a chloroethyl unit to a nucleic acid base to initiate cross linking.

Some antibiotics possess functional groups that can be involved in alkylation and cross-linking. Streptozocin **7.27** obtained from cultures of a *Streptomyces* species, is an N-nitrosourea that can methylate amino groups while mitomycin C **7.28** from another *Streptomyces* species, pos-sesses not only an aziridine ring but also a reactive urethane and a quinone.

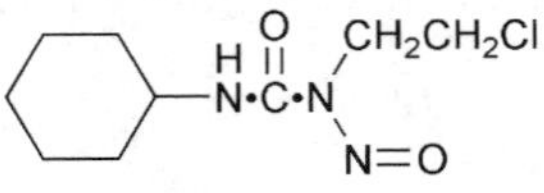

7.26 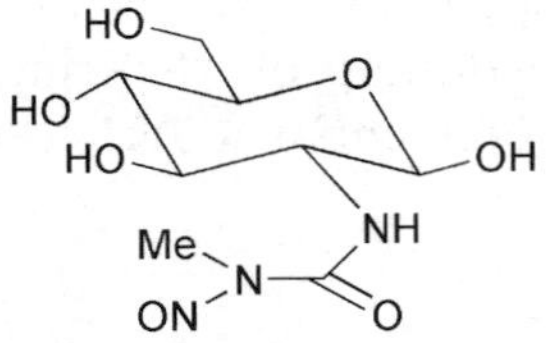   7.27

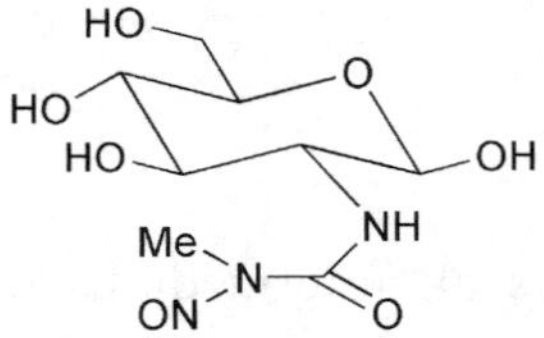

7.27 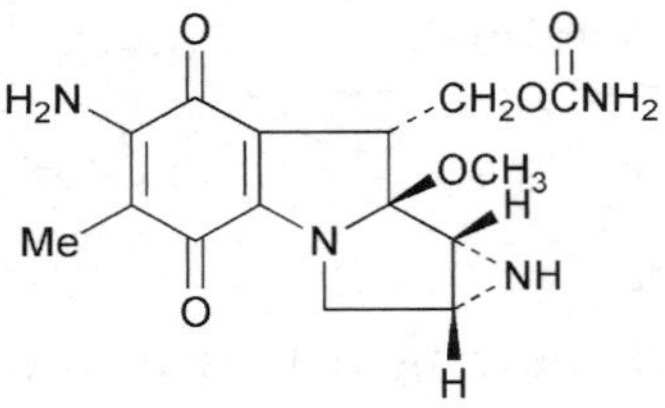   7.28

The metal complex, cisplatin (cisdichlorodiaminoplatinum II) **7.29** is another agent that forms intrastrand cross-links between adjacent gua-nine residues in DNA thus inhibiting their replication. In 1962 a phys-icist, B. Rosenberg, was investigating the way in which electric fields might interfere with the division of cultured bacterial cells. The fields were generated between platinum electrodes. The bacteria failed to grow. It was found that small amounts of platinum were dissolving to form platinum complexes with the ammonia and the chloride ions that were present in the medium, and were inhibiting bacterial cell division. This led to the development of cis-platin **7.29** as a tumour-inhibiting agent. Only the cis isomer was active. Clinical trials commenced in 1972 and the drug was launched in 1978. A number of less toxic analogues such as carboplatin **7.30** have been developed subsequently.

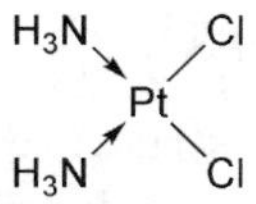

7.29

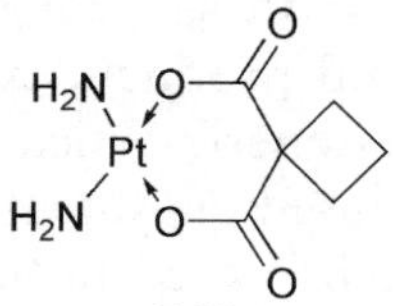

7.30

7.7 INTERCALATING AGENTS

Some antibiotics that possess a relatively planar ring system, interact with DNA by insertion into the double helix. They possess groups such as sugar units, which can hydrogen bond to the sugar-phosphate backbone of DNA. The best-known examples of this are the anthracycline antibiotics, daunorubicin **7.31** and doxorubicin **7.32**. The intercalation and binding to DNA prevents replication and can cause breaks in the strands of DNA particularly in the presence of topoisomerase II, an enzyme that alters the shape of DNA to avoid strain as it coils. An alkaloid camptothecin from a Chinese tree *Camptotheca acuminata* inhibits the function of the topoisomerase enzymes. An analogue irinotecan (Campto®) has been used, particularly in combination with fluorouracil in the treatment of bowel cancer.

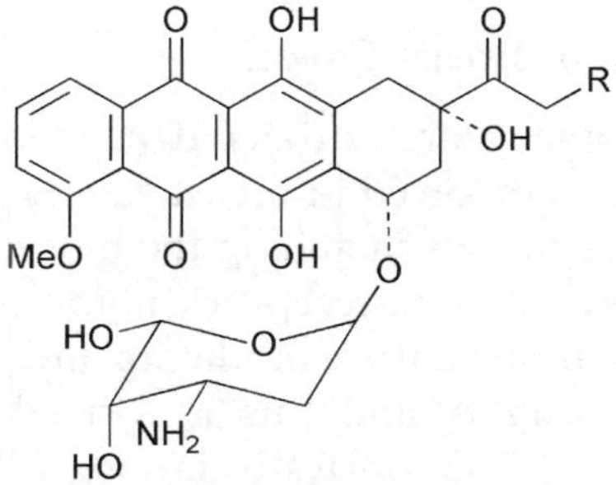

7.31 R = H
7.32 R = OH

A number of antibiotics bind to DNA and interfere with the synthesis of RNA and hence of important proteins. Bleomycin is a mixture of glycopeptides that chelate with metal ions such as copper and iron and produce breaks in the DNA. The anthramycin group of antibiotics bind to a guanine in the narrow groove of DNA.

7.8   ANTI-MITOTIC AGENTS

The mitotic spindle is a cellular structure that plays an important role in the partitioning of DNA into two daughter cells on mitosis. Microtubular proteins are a structural part of the mitotic spindle of dividing cells. Substances that bind to these proteins halt the process of cell division. A number of natural products exert their tumour-inhibitory effect this way including the dimeric Vinca indole alkaloids, vincristine and vinblastine, which are used in the treatment of leukaemia and taxol® (paclitaxel) **7.33,** which is used in the treatment of ovarian and breast cancer. Similar effects have been observed with a lignan, podophyllotoxin **7.34**.

7.33

7.34

7.9   INTERFERENCE WITH SELECTED DEVELOPMENTAL PROCESSES

7.9.1   The Treatment of Breast Cancer

A number of breast cancers are estrogen dependent. The estrogens, e.g. estrone **7.38** or estradiol involved in breast cancer are produced locally. There are a number of targets including the biosynthesis, the transport of the estrogen and the estrogen receptor. Inhibition of the biosynthesis of these steroids is an important chemotherapeutic target. The estrogens possess an aromatic ring A and this is formed from an androgen, androst-4-en-3,17-dione **7.35** or testosterone. Aromatase is the key enzyme system involved in this stage of the biosynthesis and it operates by the sequence **7.35–7.38**. The oxidation of the methyl group to form **7.36** involves a cytochrome P_{450} system. Considerable work has been done to design selective inhibitors. The iron of the cytochrome is complexed by azoles such as anastrazole (Arimadex®) **7.39** and letrazole (Femara®) **7.40**. Glutethimide **7.41** is another inhibitor, which binds to the iron of the cytochrome.

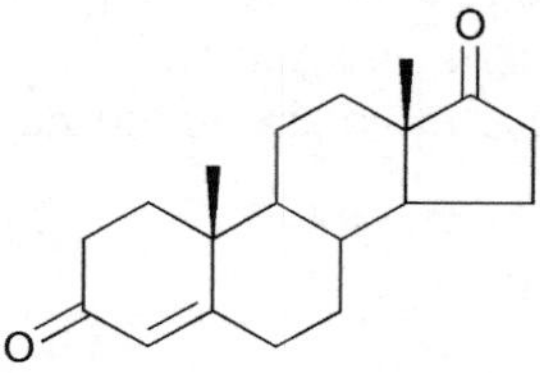

7.35

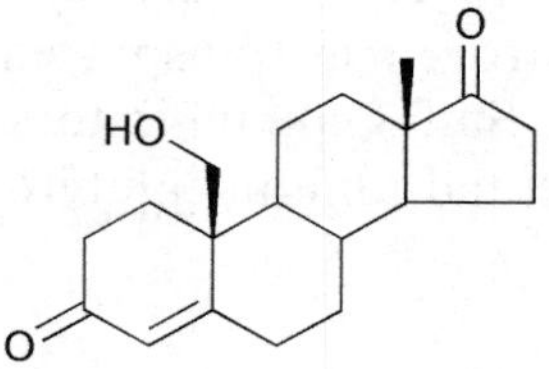

7.36

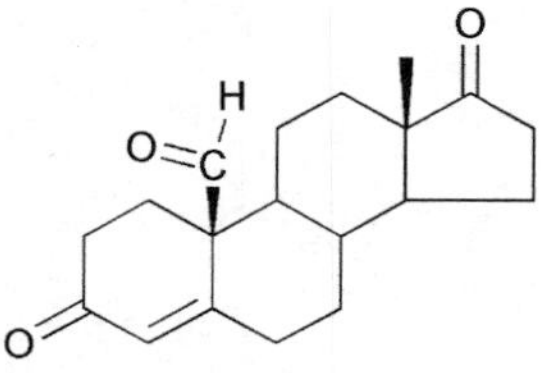

7.38

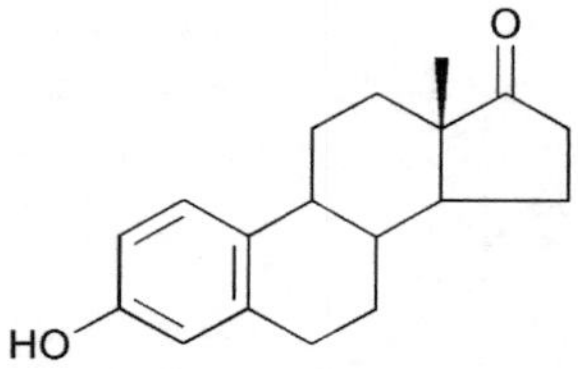

7.37

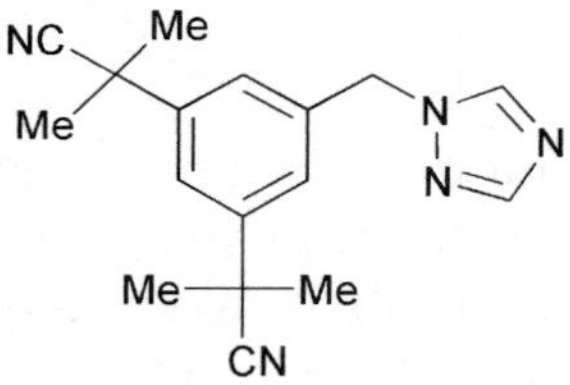

7.39

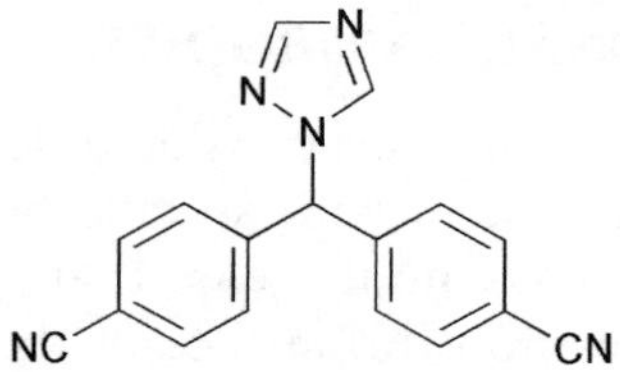

7.40

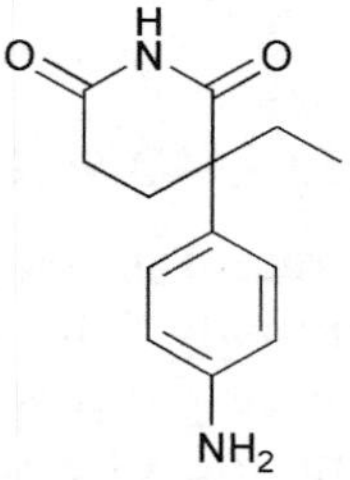

7.41

Other inhibitors of this step are steroidal analogues of the natural substrate. The most successful compounds are 4-hydroxyandrostenedione (Formestane®) **7.42** and exemestane (Aromasin®) **7.43**.

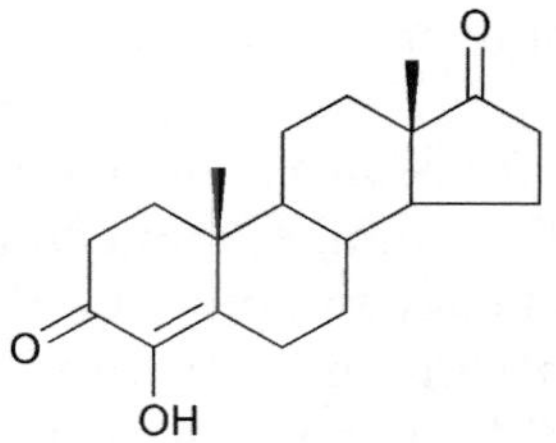

7.42

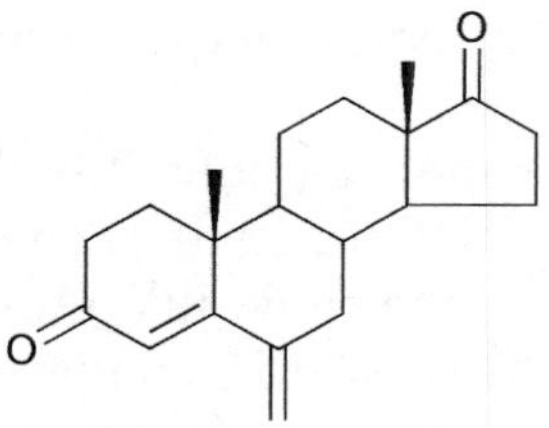

7.43

Another approach is to use compounds that are estrogen antagonists and bind to the estrogen receptor. Tamoxifen **7.44** is an example. There are two geometrical isomers of tamoxifen but only one, **7.44** has anti-estrogenic activity. It is possible to see a structural similarity to estrone in this stilbene. An active metabolite of tamoxifen is the phenol **7.45**. The estrogen is transported from its site of biosynthesis as the 3-0-sulfate. Sulfamate inhibitors of the sulfatase have shown some promise as potential tumour inhibitors.

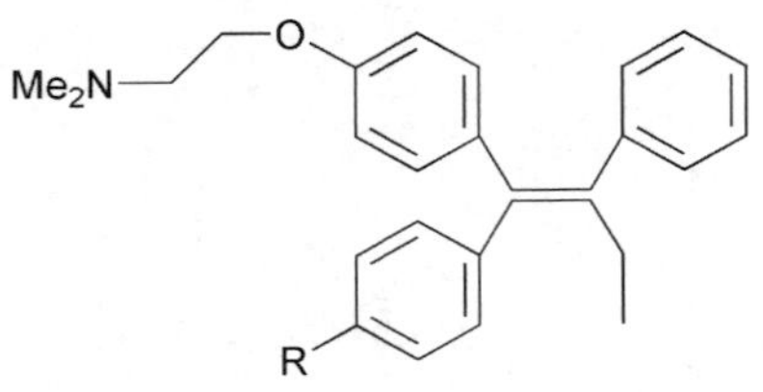

7.44 R = H
7.45 R = OH   7.46

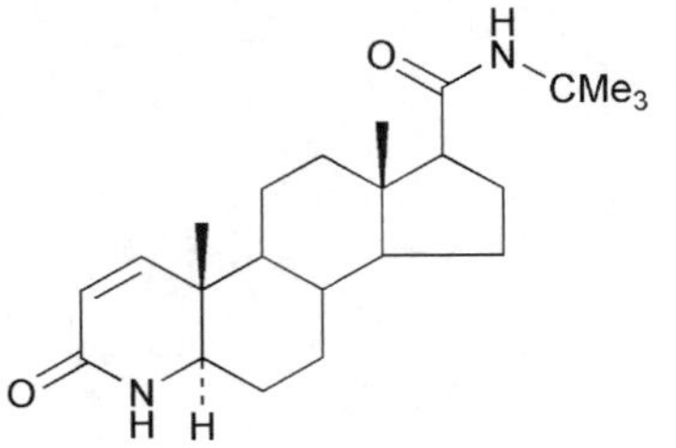

7.10   MONOCLONAL ANTIBODIES

A recent development that has considerable promise uses a monoclonal antibody. This is a protein with specific binding characteristics. Some breast cancers have tumour cells that produce abnormally large amounts of a receptor on the surface of the cancer cell. This protein is known as the human epidermal growth factor receptor 2, HER2. When its natural ligand binds to it, the growth of the cancer cell is stimulated. A monoclonal antibody that recognizes and selectively binds to the HER2 receptor has been developed. It does not bring about the binding response. This antibody known as trastuzumab (Herceptin®) by binding to the HER2 receptor on the surface of the breast cancer cells stops their growth.

Some other monoclonal antibodies are being examined in the context of other cancers. Cetuximab (Erbitux®) is another monoclonal antibody that recognizes the epidermal growth factor receptors on the surface of cancer cells. It is used in conjunction with chemotherapy in the treatment of bowel cancer. Rituximab (Mabthera®) attaches itself to a protein CD20, which is found on the surface of white blood cells (B-cell lymphocytes). This protein is present on the surface of abnormal B-cell lymphocytes, which occur in most non-Hodgkin's lymphomas. Although this antibody attacks both normal and abnormal cells, the normal cells are replaced more rapidly.

7.11 PROSTATE CANCER

Prostate enlargement (benign prostatic hyperplasia) is a serious problem for men particularly in old age, bringing about urinary problems. An enlarged prostate can become cancerous. The enlargement is brought about by the formation of excess 5α-dihydrotestosterone from the male hormone, testosterone. Hence the inhibition of testosterone 5α-reductase has been a target. A successful inhibitor is the lactam, finasteride **7.46**. This provides a means of control of this disease before it can become cancerous.

Further Reading

There are a number of longer textbooks of medicinal chemistry including the following:

G.L. Patrick, *An Introduction to Medicinal Chemistry*, 3rd edn, OUP, Oxford, 2005.

G. Thomas, *Medicinal Chemistry, an Introduction*, Wiley, Chichester, 2000.

R.B. Silverman, *Organic Chemistry of Drug Design and Drug Action*, 2nd edn, Elsevier Academic Press, New York, 2004.

T. Nogrady, *Medicinal Chemistry, a Biochemical Approach*, OUP, New York, 1988.

W.O. Foye, T.L. Lemke and D.A. Williams, *Principles of Medicinal Chemistry*, 4th edn, Williams and Wilkins, Philadelphia, 1995.

In addition there are a number of textbooks with chapters that provide case histories of the development of individual drugs. These include:

S.M. Roberts and C.R. Ganellin (eds), *Medicinal Chemistry, the Role of Organic Chemistry*, Academic Press, London, 1993.

F.D. King (ed), *Medicinal Chemistry, Principles and Practice*, The Royal Society of Chemistry, Cambridge, 1994.

J. Mann and J.C. Crabbe, *Bacteria and Antibacterial Agents*, Spektrum, Oxford, 1996.

R. Challand and R.J. Young, *Antiviral Chemotherapy*, Spektrum, Oxford, 1997.

J. Saunders, *Top Drugs*, OUP, Oxford, 2000.

Books that give an account of the development of medicinal chemistry include:

J. Mann, *Murder, Magic and Medicine*, OUP, Oxford, 1992.

J. Mann, *Life Saving Drugs, the Elusive Magic Bullet*, The Royal Society of Chemistry, Cambridge, 2004.

Comprehensive treatments of medicinal chemistry are to be found in:
C. Hansch, P.G. Sammes and J.B. Taylor (eds), *Comprehensive Medicinal Chemistry*, Pergamon Press, Oxford, 1990.
M.E. Wolff, *Burger's Medicinal Chemistry*, 5th edn, Wiley, Chichester, 1995.

Books that describe the uses of specific drugs include:

British National Formulary, 46th edn, British Medical Association, London, 2003.
J.A. Henry (ed), *New Guide to Medicines and Drugs*, 6th edn, British Medical Association, London, 2004.
M.J. Mycek, R.A. Harvey and P.C. Champe, *Pharmacology*, 2nd edn, Lippincott, Williams and Wilkins, Philadelphia, 1997.

A useful dictionary of medical terms is:

The Oxford Concise Medical Dictionary, 6th edn, OUP, Oxford, 2003.

A glossary of the terms used in medicinal chemistry has been published by IUPAC in *Pure and Applied Chemistry*, 1998, 70, 1129–1143. Many of the scientific papers describing advances in medicinal chemistry are published in the *Journal of Medicinal Chemistry, Chemical and Pharmaceutical Bulletins, Bio-organic and Medicinal Chemistry* and in more specialist journals such as the *Journal of Antibiotics*. Progress in medicinal chemistry is reviewed in the *Annual Reports of Medicinal Chemistry* (vol. 38, 2003) published by the Medicinal Chemistry Division of the American Chemical Society and Elsevier.

Glossary

ACE Inhibitors. ACE inhibitors are drugs, which are inhibitors of the angiotensin converting enzyme and are used as anti-hypertensive agents.

Acetaminophen. Acetaminophen (tylanol) is the American name for paracetamol.

Acetylcholinesterase. Acetylcholinesterase is an enzyme that is involved in the hydrolysis of acetylcholine. Acetylcholinesterase inhibitors are used as insecticides and in the treatment of Alzheimer's disease.

ADME. This is an acronym for the adsorption, distribution, metabolism and excretion of a drug.

Adrenal glands. The adrenal glands are situated just above the kidneys. There are two parts to these glands. The inner core or medulla, which produces adrenalin; and the outer cortex, which produces the corticosteroids that target the liver and the kidneys. These affect carbohydrate metabolism and electrolyte balance.

Adrenergic receptors. These are receptors for which adrenalin and noradrenalin are ligands.

Agonist and antagonist. An agonist is a substance that binds to the same receptor as a natural ligand and produces a similar biological effect. A partial agonist binds to the receptor producing a limited biological effect. An antagonist binds to the receptor but does not produce the biological effect. It may oppose the action of an agonist.

Allosteric binding site. An allosteric binding site on a receptor is different from the normal binding site for the natural ligand. When a drug binds to an allosteric binding site, it modifies the shape and action of the receptor and can affect the binding of the normal ligand.

Alzheimer's disease. A neurodegenerative disease characterized by the loss of memory (dementia). It is associated with a decrease in acetylcholine levels in the brain and with the formation of β-amyloid plagues and neurofibrilliary tangles.

Aminoacyl transferRNA. The aminoacyl transferRNA complexes carry the amino acids to the ribosomes for protein synthesis.

β-Amyloid plaques. β-Amyloid plaques are deposits of a glycoprotein found in the brain of patients with Alzheimer's disease.

Anaesthetic. An anaesthetic is a compound that reduces or abolishes sensation. A general anaesthetic affects the whole body, while a local anaesthetic affects a particular area.

Analgesic. Analgesics are drugs, which relieve pain.

Angina. Angina is a chest pain, which occurs when insufficient oxygen reaches the heart muscle.

Angiogenesis. Angiogenesis is the formation of new blood vessels to supply a group of cells.

Antigen. An antigen is a substance that binds specifically to an antibody. Allergens such as pollen, feathers or dust are antigens that cause an over-reaction in sensitive people.

Apoptosis. The death of a cell can proceed by apoptosis or by necrosis. In the former, the cell is dismantled in an ordered manner following defined biochemical steps, while in the latter the dying cell bursts apart scattering its contents and damaging adjacent cells.

Arteries. In the circulatory system, the arteries carry the blood away from the heart. The main artery adjacent to the heart is the aorta.

Asthma. Asthma is a variable constriction in the airways, which produces difficulty in breathing during an attack.

Bacterial meningitis. Bacterial meningitis leads to an inflammation of the meninges, which is the connective tissue that lines the skull and the spine enclosing the brain and the spinal cord. It produces a high fever and can be fatal.

Benign prostatic hyperplasia. Benign prostatic hyperplasia is the enlargement of the prostate gland in men. It leads to pressure on the urinary system and to difficulties in passing urine. It can become cancerous.

β-Blockers. These are drugs that are used in the treatment of heart disease and function by acting as antagonists of the β-adrenergic receptor.

Bile acids. These are steroidal cholic acids, which are produced by the gall bladder and facilitate the adsorption of fats from the intestine.

Bioisostere. Bioisosteres are structural units of comparable size that are used interchangeably without significantly changing the overall biological activity of a drug. A bioisosteric replacement may be used in 'fine-tuning' the structure:activity relationships in a series of drugs.

Carboxypeptidase. Carboxypeptidases are enzymes that hydrolyse the terminal peptide bond at the carboxyl end of the chain.

Cerebral cortex. The Cerebral cortex is the highly folded outer layer of the brain and is the part of the brain associated with perception, memory and thought and with initiating voluntary activity.

Cholesterol. Cholesterol is a sterol that forms a major lipid component along with fatty acids in membranes in man. It is transported in the circulatory system in association with a protein, a lipoprotein. The formation of plaques containing cholesterol and this lipoprotein can limit blood circulation and lead to circulatory disease.

Cholinergic receptors. The cholinergic receptors are those for which acetylcholine is a ligand.

CNS. CNS is the abbreviation for the central nervous system (brain and spinal cord) as opposed to the peripheral nervous system.

Co-enzyme. A co-enzyme is a compound, which is bound to the active site of an enzyme and which is essential for mediating the enzyme's action.

Corticosteroids. The corticosteroids ae steroid hormones that are produced by the adrenal glands. They have two main types of activity. Their glucocorticoid activity stimulates the utilization of carbohydrate, fat and protein and mediates a normal response to stress, while the mineralcorticoid activity regulates salt and water balance.

Cytochromes. The cytochromes are proteins which contain haem and in which a reversible oxidation and reduction of the central iron atom can occur. They mediate the insertion of oxygen into organic molecules. The cytochromes have a characteristic ultraviolet absorption. The cytochrome P450S develop strong absorption at 450 run in the presence of carbon monoxide.

Cytokines. The cytokines are proteins such as the interleukins and interferons which are released by cells when they are activated by an antigen. They mediate the immune response by interactions with specific receptors on the cell surface of leucocytes.

Depression. Clinical depression is characterized by excessive sadness. The activity of the patient can be agitated and restless or it may be slow. Behaviour is governed by pessimism and despair.

Diuretics. Diuretics are compounds that target the kidneys and increase the volume of urine that is produced.

DNA gyrase. DNA gyrase is an enzyme system that is involved in the coiling of DNA.

Endocrine glands. The endocrine glands are those glands that produce hormones. They include the pituitary, thyroid and andrenal glands, the ovaries, the testes, the placenta and part of the pancreas.

Endorphins. A group of peptides found in the brain, which bind to the opioid receptors. They are related to the enkaphilins.

Endothelium. The endothelium is a layer of cells that line the surface of blood vessels, the heart and the lymphatic system.

Epilepsy. Epilepsy is a disorder of the function of the brain. It is characterized by recurrent seizures that have a sudden onset.

Epinephrine. Epinephrine is the American name for adrenalin. Norepinephrine is used for noradrenalin.

Erythrocytes. Erythrocytes are the red blood cells, which contain haemoglobin.

Estrogens. The estrogens and progestogens are steroidal female sex hormones. The estrogens promote the growth of the female sex organs and female cyclical characteristics, while the progestogens maintain the normal course of a pregnancy.

Eukaryote and prokaryotes. A eukaryote is a multi-cellular organism such as a plant or an animal in which the cells contain a discrete nucleus as opposed to a prokaryote. The latter are single-celled organisms such as bacteria that do not contain a discrete nucleus.

Gall bladder. The gall bladder lies underneath the liver and excretes the bile acids into the duodenum. Absorption of food takes place from this part of the small intestine. The bile acids facilitate this process.

Genome. The human genome is the total genetic material of man. Genomics is the study of the total genetic information contained in the nucleic acids.

Glaucoma. Glaucoma is a condition that can lead to blindness and in which there is an abnormally high pressure in the ocular fluid in the eye.

G-protein. A G-protein is a protein, which binds a guanine nucleotide. The binding of a hormone to its trans-membrane protein receptor can promote the exchange of a bound GDP for GTP on a G-protein.

Gram's test. In the Gram's test, a film of the bacteria are stained with a violet dye. They are then treated with alcohol and stained a second time with a red dye. Gram-positive bacteria retain the initial stain and appear violet, while Gram-negative bacteria lose the violet stain in the alcohol but take up the red dye and so appear red.

Herpes. The herpes virus produces an inflammation of the skin and mucous membranes that leads to collections of small blisters such as cold sores around the lips.

Hippocampus. The hippocampus is an area of the brain involved in the physical aspects of behaviour governed by emotion.

Hormones. Hormones are chemical messengers, which are produced by one cell and are translocated to a receptor to produce their biological effect. An example is adrenalin.

Hypertension. Hypertension describes the elevation of blood pressure above the normal range.

Hypnotics. Hypnotics are drugs which induce sleep.

Hypothalamus. The hypothalamus is a region of the forebrain, which is linked to the thalamus and the pituitary gland. It functions as a centre for regulating various aspects of hormonal activity. It produces a series of releasing hormones, which target the pituitary gland. This small gland at the base of the skull secretes a series of peptide trophic hormones such as the adrenocorticotrophic hormone that target particular glands, for example the adrenal glands.

Immune system. The immune system involves the production of antibodies and white blood cells, which protect the body against invasive foreign cells. The organs that are responsible for the formation of these are part of the lymphatic system.

Immunosuppresants. An immunosuppresant is a drug, which inhibits the immune response. Immunosuppresants are used to reduce the chances of rejection in organ transplants.

Interferons. Interferons are proteins that are produced naturally and which act against viral infection.

Isoproterenol. Isoproterenol is the American name for isoprenaline.

Kinase. A kinase is an enzyme system that phosphorylates a hydroxyl group and may activate an enzyme system.

Log P. The logarithm of the octanol:water partition co-efficient for a drug. It reflects its hydrophobic character.

Lymphatic system. The lymphatic system is a network of capillary vessels that convey water, electrolytes and proteins from tissue fluids to the blood stream. At various points in the lymphatic system, there are lymph nodes, which act as filters preventing foreign particles from entering the blood stream. Metastases can develop at these sites.

Lysis. Lysis involves the cleavage particularly of the cell wall allowing the cellular components to escape.

Malaria. Malaria is an infectious disease, which is caused by the presence of a protozoan of the genus, *Plasmodium*, in the red blood cells. It is transmitted by the bite of an *Anophales* mosquito.

M and B. May and Baker, the company at Dagenham in Essex, which made the sulfonamide M and B 693.

Malignant. Malignant cancers are those, which are spreading to other parts of the body through the formation of metastases.

Mitosis. Mitosis is the process of cell division.

Monoamine oxidases (MAO). The monoamine oxidases A and B are enzyme systems that oxidize an amine to an imine, which is then hydrolysed to a carbonyl compound.

Mycobacterium tuberculosis. The disease is characterized by the formation of nodular lesions in the tissues of the lung and lymph glands.

Narcosis. Narcosis refers to a state of diminished consciousness.

Neurofibrilliary tangles. Neurofibrilliary tangles are microscopic threads of cytoplasm arising from nerves, which have been disorganized. These are also found in the brains of patients with Alzheimer's disease.

Neuroleptic agents. Neuroleptic agents are tranquillizing, anti-psychotic drugs.

Neuromuscular action. Neuromuscular action involves the interaction between a nerve and a muscle.

Ovary. The two ovaries are female reproductive organs situated in the lower abdomen. They produce the ova (egg cells) and a group of steroid hormones. The follicles within which the ova develop produce the estrogens. After ovulation a corpus luteum forms, which secretes progesterone to facilitate the maintenance of any pregnancy which may occur.

Pancreas. The pancreas is a gland just behind the stomach, which secretes a mixture of digestive enzymes into the duodenum.

Parietal cells. The parietal (oxyntic) cells are found in the gastric glands in the fundic region of the stomach and release acid to aid the digestive process.

Parkinson's disease. Parkinson's disease is characterized by a tremor, a slowness of movement, and a rigid stooping posture. It is a degenerative disorder of cells in the brain associated with ageing.

Peptic ulcer. A peptic ulcer is a lesion in the mucosal lining of the digestive tract, which is reluctant to heal.

Pharmacophore. The part of the molecule that is responsible for the biological activity.

Pharmacokinetics. Pharmacokinetics involves the study of those features, which affect the absorption, distribution, metabolism and excretion of a drug.

Placebo. An inactive substitute for an active drug, which is used in clinical trials.

Platelets. Platelets are microscopic disc-shaped structures present in the blood. Platelets, by aggregating, play an important role in blood coagulation and the healing of wounds.

Pro-drug. A compound, which is converted into the active drug in the body.

Protozoa. Protozoa are a group of single-cell organisms some of which are parasitic in man.

Psoriasis. Psoriasis is a chronic skin disease in which pink scaly patches form on the skin particularly on the elbows and the scalp.

Psychotropic agents. Psychotropic agents are drugs that affect mood.

QSAR. QSAR is an abbreviation for quantitative structure:activity relationships in which a relationship is sought between a physicochemical parameter such as pK and the biological activity.

Rhinitis. Rhinitis is an inflammation of the mucous membranes of the nose leading to excessive secretions of mucous.

Ribosome. The ribosome is a cellular structure that comprises RNA and protein and is a site of protein synthesis in the cell. The bacterial system contains two unequal sub-units. The 30S and 50S sub-units are parts of this structure. The S stands for Svedberg unit and is related to the molecular weight of the sub-unit and its centrifugation.

Rickets. Rickets in children is a disease in which the bones do not harden to make rigid structures. Long bones as in the legs become bowed and the ribcage becomes deformed.

Schizophrenia. Schizophrenia is a severe mental disorder characterized by delusions, hallucinations and a loss of contact with reality.

Second messenger. Second messengers are hormones such as cyclic adenosine monophosphate (cAMP), which are produced within the cell as a result of an external stimulus. The second messengers target sites within the cell.

Septicaemia. Septicaemia involves the bacterial destruction of tissue and the discharge of bacterial toxins into the blood.

Sporozoites. Sporozoites and merozoites represent developmental stages in the life cycle of protozoa. The organism in these various stages are known as schizonts.

Synapse. The synapse or synaptic cleft refers to the narrow gap between a nerve ending and the target cell.

Syphilis. Syphilis is a sexually transmitted disease caused by a bacterium, *Treponema pallidum.* It produces lesions, fever and can eventually lead to general paralysis and death.

Teratogen. A teratogen is a substance that induces the formation of developmental abnormalities in a fetus and leads to the birth of a child with deformaties.

Testes. The testes are a pair of male sex organs, which produce the spermatozoa and secrete the androgens. These steroid hormones regulate male secondary characteristics.

Thrombosis. Thrombosis is a condition in which the blood forms a clot (a thrombus).

Thyroid gland. The thyroid gland is situated at the base of the neck. There are two lobes either side of the trachea. The thyroid hormones, triiodothyronine and thyroxine regulate many normal metabolic processes.

Transition state inhibitor. Transition state inhibitors are enzyme inhibitors that are designed to mimic the transition state in an enzyme reaction.

Transmembrane protein. Transmembrane proteins cross the cell membrane and are involved in signal transduction. A ligand such as a neurotransmitter binds to a receptor component of the transmembrane protein and this modifies the binding of proteins on the other side of the membrane or leads to the opening of an ion-channel.

Trypanosomiasis. Trypanosomiasis is a disease (sleeping sickness) caused by the infection of the blood by a trypanosome parasite.

Typhoid. Typhoid is a bacterial infection of the digestive system, which can lead to a high fever, inflammation of the spleen and erosion of the intestinal cell wall. The disease is transmitted through contaminated food or drinking water.

Tuberculosis. Tuberculosis is an infectious disease caused by the bacterium.

Vasodilation. Vasodilation is an increase in the diameter of blood vessels, especially the arteries. It brings about a lowering of the blood pressure. The opposite effect is vasoconstriction.

Subject Index